Anderson Assaid Simao
Angelita Duarte Correa

Medicinal Plants Used to Treat Obesity

Anderson Assaid Simao
Angelita Duarte Correa

Medicinal Plants Used to Treat Obesity

Chemical composition, efficacy and toxicity

ScienciaScripts

Cover image: www.ingimage.com

This book is a translation from the original published under ISBN 978-613-9-61375-5.

Publisher:
Sciencia Scripts
is a trademark of
Dodo Books Indian Ocean Ltd. and OmniScriptum S.R.L publishing group

120 High Road, East Finchley, London, N2 9ED, United Kingdom
Str. Armeneasca 28/1, office 1, Chisinau MD-2012, Republic of Moldova, Europe
Printed at: see last page
ISBN: 978-620-7-65482-6

Table of contents:

To the memory of my dear father, for all he could do for me, for always showing me the right way.

To my beautiful little Caua, for bringing me so much joy and motivation.

To my siblings, Andréia and Luís Fernando, for their friendship and support.

To my mother, for always trusting me and loving me.

DEDICATE

CHAPTER I

PRESENTATION

The results of this thesis are presented in the form of articles, which can be found in the articles section.

The bibliographical references refer only to the citations that appear in the introduction and theoretical references.

Each article is structured according to the standards of the scientific journals chosen for its presentation or publication.

1 INTRODUCTION

The use of medicinal plants in the prevention and treatment of diseases is a very old practice and has a historical importance in the transformation of therapeutics, both as a source of pharmaceutical raw materials and as a source of active substances isolated and used as prototype medicines and, more recently, in the form of herbal medicines.

The growing number of diseases and the lack of options, access, safety and efficacy of treatments for these ailments have led many people to look to medicinal plants as a way of combating them. An important example of this practice is the treatment of obesity, one of the main problems facing the health sector today and which, due to the proportions it has reached, has mobilized great efforts in the production of knowledge. Its incidence is independent of socio-economic factors and age and its consequences range from the development of debilitating diseases to death, directly affecting individuals' quality of life.

It is estimated that there are more than one billion overweight adults in the world, and of these, 400 million are clinically obese (TUCCI; BOYLAND; HALFORD, 2010), which has led to the disease becoming a global epidemic. In Brazil, according to the Ministry of Health, approximately 50% of the population is overweight and 15.8% is obese, and estimates suggest that, by 2025, Brazil will be the fifth most obese country in the world (BRASIL, 2012b).

Obesity causes psychological problems, frustration and predisposes the body to a series of diseases, including cardiovascular disease, some types of cancer, diabetes and high blood pressure, causing economic costs for both the state and society.

There is a consensus among the scientific community, the medical profession and various international health organizations that the main causes of obesity are: an inadequate diet, rich in carbohydrates and saturated fats, a sedentary lifestyle and a lack of regular physical activity. In addition to the health issue, there is also the question of aesthetics, which has been highly valued in recent years and which has led many people to look for options for treating obesity.

Of the options available for treating this disease, the most commonly used are balanced diets, regular exercise and drug treatments, ranging from lipase inhibitors to anorexiants.

Due to the side effects, ineffectiveness and high cost of the drugs traditionally used to treat obesity, the use of medicinal plants has been widely explored, both by the population, due to their easy access, low cost, non-requirement of a medical prescription and belief in the absence of toxic effects, and by the pharmaceutical industry, which sees these plants as a viable alternative for the future development of drugs that induce weight reduction effectively and safely (MAYER et al., 2009; PARK; LEE; SUNG, 2005). Studies show that various natural products, including extracts and compounds isolated from plants, are being used to reduce body weight and prevent obesity (RAYALAM; DELLA-FERA; BAILE, 2008; SIMAO; CORREA; CHAGAS, 2012; SOUZA et al., 2011).

The importance of medicinal plants is due to their contribution as a natural source of drugs and because they offer great possibilities for obtaining prototype molecules due to the great diversity of their constituents (YUNES; CALIXTO, 2001). However, many plants are used in herbal preparations without proper quality and safety control, as scientific literature shows that most medicinal plants contain toxic substances and/or variable chemical composition (CAPASSO et al., 2000).

In addition, some pharmaceutical companies, whose sole aim is to make a profit, proclaim the existence of miracle products, advertising them in a misleading way and, in most cases, without concern for public health, offering so-called natural products, but which, in most cases, have undeclared substances added to their formulas and, in some cases, there is a lack of studies on the constituents present in the formulas.

An important example of this practice is the use of the herbal medicine Moder diet, made up of a combination of extracts from the plants *Aloe vera* (L.) Burm. (aloe), *Simaba ferruginea* St. Hil. (calunga), *Baccharis trimera (*Less.) DC (carqueja), *Garcinia cambogia* Desr. and *Tournefortia paniculata* Cham. (marmalade), which have been freely marketed in Brazil, with the supposed effect of appetite suppressant and weight reducer, without there being any scientific evidence to guarantee the safety of their use. However, the National Health Surveillance Agency (ANVISA) has banned the sale of this herbal medicine due to the lack of studies attesting to its efficacy and the absence of risks during its use (AGÊNCIA NACIONAL DE VIGILÂNCIA SANITÁRIA - ANVISA, 2009).

Recently, the use of the Moder diet has experienced a boom, reflected in the emergence of countless cases of weight loss and its dissemination mainly on websites and in gyms. However, due to the surprising effects provided by this slimming agent, there is concern about the possibility of adulteration, especially by anorexics, and the toxicological risks arising from continuous use, highlighting the importance of studies with the plants that are part of its composition to prove its therapeutic and toxicological effects. In view of the search for lower-cost, easily accessible and proven safe therapeutic alternatives for the treatment of obesity, as well as the lack of scientific information on the Moder diet herbal medicine and the plants that make it up, a detailed investigation into its chemical constituents, possible toxicity resulting from continuous use and elucidation of the mechanism of action is necessary.

1.1 General objective

Determine the chemical composition of the plants that make up the Moder diet herbal medicine, carry out digestive enzyme inhibition tests, measure the antioxidant potential and biological tests with the plant that shows the greatest presence of phytochemicals with pharmacological properties, with a view to its therapeutic (potential for preventing and treating obesity) and toxicological characterization.

1.2 Specific objectives

a) Carry out phytochemical screening, determine the centesimal and mineral composition and the content of some bioactive compounds of the plants that make up the phytotherapy of the Moder diet.

b) Carry out digestive enzyme inhibition tests in the presence of simulated gastric fluid with the plants that make up the herbal medicine Moder diet and with the mixture equivalent to the herbal medicine.

c) To evaluate the antioxidant potential of the plants that make up the Moder diet herbal medicine and the mixture equivalent to the herbal medicine.

To evaluate the anti-obesity effects of the plant with the highest presence of phytochemicals with anti-obesity properties, inhibition of digestive enzymes and antioxidant activity, in *Wistar* rats subjected to a hypercaloric diet, using the following parameters: body weight, food consumption, glycemia, lipids, lipid peroxidation and toxicity (genotoxicity and histology).

2 THEORETICAL FRAMEWORK

2.1 Epidemiology: relevance of the problem

Due to the consequences caused by obesity and the speed with which it spreads around the world, it has been considered a global epidemic, with more than one billion overweight adults, 400 million of whom are clinically obese, affecting both developed and developing countries, including Brazil (TUCCI; BOYLAND; HALFORD, 2010; WORLD HEALTH ORGANIZATION - WHO, 2010). In America, obesity has increased in both sexes, in both developed and developing countries. In Europe, there has been an increase in obesity of between 10% and 40% in most countries over the last decade. Some countries in the Western Pacific region, such as Australia, Japan, Samoa and China, have also seen an increase in the prevalence of obesity. On the African continent, obesity is still relatively uncommon, and its prevalence is higher in the urban population than in the rural population (WANDERLEY; FERREIRA, 2010).

In Brazil, the analysis of five population surveys carried out by the Ministry of Health - the

National Household Expenditure Study (ENDEF), carried out between 1974-1975; the National Health and Nutrition Survey (PNSN), in 1989; the Living Standards Survey (PPV), in 1996-1997; and the Household Budget Survey (POF), in 2002-2003 and 2008-2009 - made it possible to assess the emergence of obesity and verify its main determinants. According to these studies, obesity and overweight are increasing rapidly in the Brazilian population, especially among adults, while the prevalence of malnutrition has decreased rapidly in recent decades. Data from the Brazilian Institute of Geography and Statistics (IBGE) show that excess weight in the male population aged 20 and over jumped from 18.5% in 1974-1975 to 50.1% in 2008-2009. In women, the increase was from 28.7% to 48% in the same period (INSTITUTO BRASILEIRO DE GEOGRAFIA E ESTATÍSTICA - IBGE, 2009).

With regard to children and adolescents, the data on the rise in obesity is equally worrying. To put the obesity epidemic among children in the United States into perspective, the number of obese children exceeds the combined number of children affected by the AIDS virus, cancer, childhood diabetes and cystic fibrosis (FREEDMAN et al., 2007). In the United Kingdom, 16% of children are obese and another 14% are overweight.

In Brazil, overweight in this section of the population is growing rapidly: in 1974, overweight affected 4.9% of children aged 6 to 9 and 3.7% of adolescents aged 10 to 18. Between 1996 and 1997, there was a 14% overweight rate in the 6 to 18 age group (WANG et al., 2002). Data released by the POF (2008-2009) indicates that 33.5% of Brazilian children aged 5 to 9 are overweight, with 16.6% of all boys being obese and 11.8% of girls being overweight (IBGE, 2009). The prevalence of obesity in this age group is worrying due to the high risk of obese children becoming obese adults, with various associated morbid conditions.

Recent studies show that almost half of the Brazilian population is overweight. The results of the latest Surveillance of Risk and Protective Factors for Chronic Diseases by Telephone Survey (VIGITEL), promoted by the Ministry of Health, show that overweight and obesity have increased over the last six years in Brazil. According to the study, the proportion of overweight people in Brazil rose from 42.7% in 2006 to 48.5% in 2011. In the same period, the percentage of obese people rose from 11.4% to 15.8%. The study also shows that the increase in the percentage of obese and overweight people affects both the male and female populations. In 2006, 47.2% of men and 38.5% of women were over the ideal weight. Now, the proportions have risen to 52.6% and 44.7% respectively (BRASIL, 2012b).

WHO estimates suggest that by 2015, the number of obese and overweight people in the world will more than double compared to current figures, reaching 700 million and 2.3 billion, respectively. For Brazil (age group 15 to 100), it is estimated that 66.5% of women and 60.3% of men will be overweight or obese (WHO, 2010), highlighting the urgent need for safe and effective alternatives to curb the spread of this disease.

2.2 Obesity: concept and classification

Obesity can be generally defined as a disease resulting from the abnormal or excessive accumulation of fat in the form of adipose tissue, which is the consequence of a positive energy balance and which results in damage to health and can lead to death (SOCIEDADE BRASILEIRA DE ENDOCRINOLIGA E METABOLOGIA - SBEM, 2010).

The anthropometric measurements commonly used to diagnose obesity are the body mass index (BMI), the abdominal circumference measurement and the waist-to-hip ratio, the former being the most widely used (WHO, 2010). BMI is calculated by dividing an individual's weight (in kg) by the square of their height (in meters) (SBEM, 2010). It is the standard used by the WHO, which identifies normal weight when the result of the BMI calculation is between 18.5 and 24.9 kg/m^2 . The BMI range between 25 and 29.9 kg/m^2 is called overweight. BMI values above 30 kg/m^2 indicate obesity. Obesity is classified into three different degrees. From 30 to 34.9 kg/m^2 is grade I obesity, from 35 to 39.9 kg/m^2 is grade II obesity and above 40 kg/m^2 is grade III obesity (morbid obesity).

Overweight and obesity are risk factors for a number of diseases. When compared to normal weight individuals, obese people are more likely to develop problems such as hypertension, cardiovascular disease, type 2 diabetes, among others (SBEM, 2010; WHO, 2010).

In addition to the degree of fat, another factor that interferes with the risks associated with excess weight is its regional distribution in the body. When excess fat occurs in the abdominal region (android obesity), the risks associated with other diseases are much greater than when the fat is distributed evenly and peripherally (gynecoid obesity) (WHO, 2010).

2.3 Etiology

The etiology of obesity is not entirely clear, but there is a consensus in the literature that it is not characterized as a singular disorder, but rather as a heterogeneous group of conditions with multiple causes, which end up being reflected in the obese phenotype (ASSOCIAÇÃO BRASILEIRA PARA O ESTUDO DA OBESIDADE E DA SÍNDRO METMEABÓLICA - ABESO, 2009).

There are many causes that can lead to obesity, including genetic, environmental, social, economic and cultural factors and endocrine disorders (SBEM, 2010). Also associated with weight gain are factors such as changes at certain times in life (marriage, separation, widowhood); certain situations of violence; psychological factors (depression, anxiety and stress); drug treatments (with psychotropic drugs and corticosteroids); stopping smoking; drastic reduction in physical exercise, among others (BRASIL, 2006).

Among the various causes of obesity, the most immediate determinant of excessive fat accumulation is the positive energy balance, which occurs when the amount of energy consumed is greater than the amount expended (HENRY; PANDIT, 2009). This is fully favored by current diets, which are high in energy density, rich in saturated fats, simple sugars and low in complex carbohydrates and fiber, and by the increase in physical inactivity that makes up the contemporary Western lifestyle (POPKIN; GORDON-LARSEN, 2004).

The explanations for the rise in obesity have been somewhat controversial. Some studies have shown the existence of genetic differences between obese and non-obese people, while other studies focus on lifestyle factors such as poor eating habits and physical inactivity (PERUSSE; BOUCHARD, 1999; SUNDQUIST; JOHANSSON, 1998).

2.4 Obesity and its co-morbidities

Obesity is associated with some of the most prevalent diseases in modern society. Excess body fat increases the risk of developing diabetes, coronary heart disease, cancer, hypertension, dyslipidemia, diseases of the digestive system, lung diseases, orthopedic complications, among others, and can even lead to death (DANIELS, 2009; GUH et al., 2009).

Obesity also affects children and adolescents, who are increasingly susceptible to excess weight, with childhood obesity being a factor in adult obesity. In this age group, obesity is associated with chronic diseases such as hypertension and diabetes, as well as dyslipidemia. Other associated co-morbidities are orthopaedic problems, fungal skin infections and hepatic steatosis (DANIELS, 2009).

Being overweight is associated with increased morbidity and mortality, and this risk increases with weight gain. Compared to normal-weight individuals, obese people are 20% more likely to die from all causes; they have double the risk of dying from diabetes; they are 40% more likely to develop gallbladder dysfunction and 25% more likely to develop coronary heart disease. In men who are 40% overweight (grade II obesity), mortality from all causes is 55% higher, they are 70% more likely to develop coronary heart disease and the risk of death from diabetes is four times higher than in people of normal weight (BLUMENKRANTZ, 1997).

2.5 Economic impacts

Research shows that there are more than a billion overweight adults in the world (TUCCI; BOYLAND; HALFORD, 2010), and this situation implies high economic costs for both government agencies and the individuals affected. In the United States, the country with the highest number of obese people in the world, a *Centers for Disease Control and Prevention* (CDC) survey carried out between 1998 and 2006 revealed that the costs of obesity increased from 6.5% to 9.2% (CENTERS FOR DISEASE CONTROL AND PREVENTION - CDC, 2012). In developing countries, the costs range from 2 to 6% of total health costs. If we take into account that an obese person develops a series of other diseases as a result of being overweight, for example: type 2 diabetes, cardiovascular diseases, among others, the true costs of this epidemic could be much higher.

In Brazil, direct spending on this disease, including interventions, consultations and medication, exceeds one billion reais per year, which is equivalent to 12% of the total annual spending on interventions by the Unified Health System (SUS). Worldwide, Brazil ranks sixth among countries with obesity problems (GIGANTE et al., 2009).

2.6 Treatment

Obesity must be recognized as a disease and treated as such. Its treatment should be complex and multidisciplinary and its choice should be based on the severity of the problem and the presence of associated complications.

There are various options for treating overweight and obesity, but fundamentally the ways forward should involve reducing energy intake and increasing calorie expenditure.

Most treatments involve balanced diets, regular exercise, drug therapy and bariatric surgery.

Drug treatment includes various agents that promote weight loss. These are divided into three categories: those that reduce hunger or modify satiety; those that reduce the digestion and absorption of food and those that increase energy expenditure (NONINO-BORGES et al., 2006).

In Brazil, there are some drugs registered for the treatment of obesity, such as sibutramine, diethylpropion (anfepramone), femproporex, mazindol and orlistat. They are only indicated in cases where the BMI is greater than or equal to 30, or greater than or equal to 25 accompanied by risk factors in cases where treatment with diets, physical exercise and behavioral changes have had no effect. However, there is a lot of controversy about its use, due to the scarcity of studies on its long-term effects and its side effects, such as heart disease, hypertension, sexual dysfunction, among others. Another issue associated with drug treatment is that, unfortunately, most of the time treatment only brings short-term benefits, which are often associated with weight gain after stopping the medication (MAHAN; SCOTT-STUMP, 2008).

Due to the seriousness of obesity and the side effects of drug treatment, the use of natural products has grown rapidly in recent years, mainly because people believe that this type of treatment has no side effects, does not require a prescription and is inexpensive.

Some of the natural products used are aloin (*Aloe vera*), artichoke (*Cynara scolynmus),* boldo (*Peumus boldus),* carqueja (*Baccharis* sp.), sacred cardamom (*Rhamnus purshiana),* centella asiatica, citrus extract (*Garcinia* sp.), chlorella (*Chorella pyrenoidosa), hawthorn* (*Maytenus ilicifolia),* spirulina (*Spirulina maxima*), *Fucus* sp, guarana (*Paullinia cupana),* false ginseng (*Pfaffia paniculata),* glucomannan (*Amorphophallus konjac),* jurubeba (*Solanum paniculatum)*, passion fruit (*Passiflora alata)* and senna (*Cassia angustifolia).* Their use has also been stimulated by advertisements in various media (internet, non-scientific lay magazines, among others), as well as many compounding pharmacies that have marketed these plants in capsule form.

Another very common practice used by the population and the herbal medicine industry is the use of a combination of plants to achieve a greater variety of effects. However, most of these products lack scientific proof of their efficacy and safety and require further study.

Plants can help treat obesity, both directly and indirectly. Directly, they can stimulate metabolism and modulate appetite. Indirectly, they can help treat obesity associated with other pathologies. In this case, we can mention plants with a diuretic action and those that act as central nervous system depressants (MORO; BASILE, 2000).

The use of natural products has been gaining a lot of credit, especially after the implementation of the National Program of Medicinal Plants and Phytotherapics (PNPMF), which aims to include medicinal plants, phytotherapics and services related to phytotherapy in the Unified Health System (SUS) with safety, efficacy and quality (BRASIL, 2006).

Another advantage of using natural products is that even a modest reduction of 5 to 10% in body weight can bring health benefits (SBEM, 2010). Studies carried out in China, Canada, the United States and several European countries have shown that more than half of overweight people were able to reduce the incidence of diabetes by reducing their weight and exercising (PANAMERICAN HEALTH ORGANIZATION - PAHO, 2003). Thus, the search for and use of medicinal plants to treat obesity has increased in recent years.

Unfortunately, most of the medicinal plants used to treat obesity are used without medical advice, which is worrying since there are few or no scientific studies on many of them, which demonstrates the need for a broad characterization of these plants to verify their efficacy and safety.

2.7 Moderate dietary phytotherapy and its components

A natural slimming agent composed of herbal extracts, Moder diet is characterized by being a plant compound used to help the body lose weight and prevent obesity. It reduces appetite and anxiety, speeds up metabolism and eliminates localized fat and flab. It leads to a loss of 6 to 8 kg per month and is said to have no side effects or contraindications (WIKIPEDIA, 2010).

Its composition includes extracts of calunga (*Simaba ferruginea*), for inappetence, a tonic used to combat anemia and weakness in general; carqueja (*Baccharis trimera*), combats intestinal malfunction, is anti-diabetic, anti-rheumatic and diuretic; marmalade (*Tournefortia paniculata*), antibiotic for the urinary tract; *aloe vera*, revitalizes and fights sagging, anti-aging; *Garcinia cambogia*, fights obesity, used as an appetite suppressant and also to block the absorption and synthesis of fats.

Produced by the Erusmed laboratory, Moder diet is in the sights of ANVISA, which, by means of resolution 3.691/2009, has ordered the suspension of the manufacture, distribution, trade and use of the drug, due to the fact that the herbal medicine has no registration, nor studies proving its efficacy and/or toxicity (ANVISA, 2009).

Sold as a natural remedy in capsule form, this herbal medicine can be mixed with anorexics, thyroid hormones and other substances whose use is prohibited or only recommended on prescription. Herbal mixtures are also not recommended. They can be incompatible and cause so-called drug interactions, with unpredictable effects, thus requiring studies to scientifically prove their effects.

Among the plant species included in the Moder diet, *Aloe vera* is the most studied and is on the RENISUS (plant species with the potential to generate products of interest to the SUS (BRASIL, 2009)), ANVISA (medicinal plants with simple phytotherapics registered with ANVISA (CARVALHO et al, 2008)), LMFRS (medicinal plants with phytotherapic derivatives on the list of phytotherapics with simplified registration (Normative Instruction 5/08) (BRASIL, 2008)) and SUS (medicinal plants with phytotherapic derivatives officially registered for use in SUS (RENAME) (BRASIL, 2012a)). *Baccharis trimera* is on the RENISUS list and registered as a DV (species that can be marketed in the form of an industrialized plant drug (Resolução da Diretoria Colegiada 10/10) (BRASIL, 2010)), and *Garcinia cambogia is* on the ANVISA list (TEIXEIRA, 2013). The species *Simaba ferruginea* and *Tournefortia paniculata* are absent from all the lists and their use is based only on descriptions of popular use.

With regard to the treatment of obesity, few studies have been carried out on these plants, which reveals the need for studies that could help in the pharmacological and/or toxicological understanding of the internal use of these plants in the fight against obesity.

2.7.1 *Aloe vera*

Aloe vera (L.) Burm. (aloe *vera*) is a plant species belonging to the genus *Aloe* and the family Liliaceae, native to North Africa (BACH; LOPES, 2007). It is a plant used for various medicinal purposes, the main applications being related to skin problems such as acne, psoriasis, leprosy and burns, as well as being widely used in the cosmetics industry (VALVERDE et al., 2005). Reports from Macedonia, Egypt, China and even the Bible show that this plant was widely used in ancient times for various purposes (BACH; LOPES, 2007).

The leaves of this plant are made up of a parenchymatous tissue rich in polysaccharides (mucilage), which gives it a viscous consistency (slime). Its active ingredients are found in this mucilage and consist of organic tissues, vitamins (A, B and C), enzymes (lipases, proteases), mineral salts and amino acids, some of which are very important for humans (BOUDREAU; BELAND, 2006; HE et al., 2005). The sap obtained from its bark is rich in alloin, allantonin and anthraquinones. These substances give it healing, antibacterial, antifungal and antiviral properties (VALVERDE et al., 2005). The biological activities reported for this plant species are generally attributed to a synergistic action of the compounds it contains, rather than just one chemical substance (TALMADGE et al., 2004).

As for its toxic potential, studies carried out with aqueous extracts of *A. vera* show that at usual doses (40 mL L^{-1}), the aqueous extract of this plant does not induce mutagenicity in *Allium cepa* cells, while at doses higher than those used popularly, the extract was able to induce cytotoxicity and mutagenicity when evaluated in the same cell type. In this study, the authors also found an antimutagenic effect (paracetamol-induced mutagenicity) of *A. vera* gel on human lymphocytes using the micronucleus test (STURBELLE et al., 2010), which was also verified by Stanic (2007) in a study on the antimutagenic action of *A. vera* gel on the effects induced by ethyl methanesulphonate.

The antioxidant capacity was observed by Yun, Juan and Qiuhui (2003), who analyzed *A. vera* at various ages and found that, at four years of age, the plant showed greater antioxidant potential than at younger ages, which shows that the growth phase plays a vital role in the antioxidant composition of *A. vera* gel. Rajasekaran, Sivagnaman and Subramanian (2005), on the other hand, observed that when the alcoholic extract of *A. vera* (300 mg kg-1) was administered orally to diabetic rats, the high levels of lipid peroxidation and hydroperoxides in the tissues of these diabetic rats were reversed to almost normal levels after treatment with the extract of this plant. The treatment also resulted in a significant increase in the levels of reduced glutathione, superoxide dismutase, catalase, glutathione peroxidase and glutathione-S-transferase in the liver and kidneys of the diabetic rats, results which highlight the antioxidant properties of the *A. vera* extract.

In relation to obesity, a study carried out with the alcoholic extract of *A. vera alcoholic extract,* administered orally at concentrations of 200 and 300 mg kg^{-1} of body weight in rats on a normal diet, rats with glucose overload and rats with streptozotocin-induced hyperglycemia, showed that the extract promoted the maintenance of blood glucose by controlling the action of enzymes that metabolize carbohydrates (RAJASEKARAN et al., 2004), which was also observed by Noor et al. (2008). Mohamed (2011) and Rajasekaran et al. (2006) observed that oral administration of *A. vera* resulted in a significant decrease in cholesterol and triglyceride levels in an experiment carried out on rats, which demonstrates that the freeze-dried extract of A. *vera* can be used as an aid in the treatment of obesity.

There are *A. vera-based* products on the North American market that promise to cure diabetes, cancer and even tuberculosis. However, the anti-bacterial and anti-inflammatory properties reported for this plant have not yet been proven in humans suffering from these diseases.

According to Brazilian legislation, only cosmetics and herbal medicines for topical use can be manufactured on an industrial scale from this plant. Products for internal use in the form of food, such as juices and isotonic drinks, marketed in other countries, are not authorized in Brazil due to the lack of research proving their safety. With regard to the treatment of obesity, few studies were found.

Given the scarcity of studies on its chemical constituents and the effects of its internal use, more detailed studies could help to understand the pharmacological and/or toxic effects of the internal use of this plant in the fight against obesity. ***Baccharis trimera***

Belonging to the Asteraceae family, *Baccharis trímera* (Less.) DC, popularly known as carqueja, is a perennial plant native to the Amazon and found in tropical regions, including Brazil and other South American countries. In folk medicine, *B. trímera has been* attributed internal use functions such as stomachic, anti-rheumatic, anthelmintic, tonic for the liver and in the treatment of diabetes, gastroenteritis, anorexia, flu, colds, as well as having a digestive, diuretic and weight-reducing action. Due to its diuretic effect, it helps with weight loss and diabetes control (OLIVEIRA et al., 2005). In external use, *B. trimera* is indicated for the treatment of wounds and ulcers (VERDI et al., 2005). In the brewing industry, it can be used as a substitute for hops.

Flavonoids and terpenoids are considered the most common groups in this plant. There are also other compounds such as saponins (SOUZA et al., 2011) and essential oils (AGOSTINI et al., 2005). Of the many biological effects of *B. trimera* already reported, the following can be mentioned: antioxidant (OLIVEIRA et al., 2003), gastroprotective (BAGGIO et al., 2003), anti-ulcer (DIAS et al., 2009), hypoglycemic (BARBOSA-FILHO et al., 2005), among others.

As for its toxic effects, the main one already reported is the induction of abortion, which has been proven experimentally in animals (VERDI et al., 2005). Low toxicity and antimutagenic effects were observed by Nakasugi and Komai (1998) in methanolic fractions isolated from *B. trimera*. Baggio et al. (2003) analyzing aqueous and hydroalcoholic extracts of the leaves of *B. trimera* and

Dias et al. (2009) analyzing the lyophilized crude extract of this plant did not observe cytotoxic effects *in Wistar* rats, which was also observed by Peron et al. (2008) when evaluating bone marrow cells from *Wistar* rats treated *in vivo* via gavage with an aqueous extract of this plant at concentrations of 6.85 and 68.50 mg mL^{-1} . Pinho et al. (2010) evaluated the in *vivo* and *in vitro* mutagenicity of *B. trimera* tea, at doses of 20 g L^{-1} (usual dose) and 200 g L^{-1} (10 times more concentrated), using the *Allium cepa* L. test and the chromosome aberration test on human lymphocytes, and reported the need for moderation in its consumption.

Rodrigues et al. (2009) analyzed aqueous extracts of *B. Trimera,* administered by gavage at doses of 200, 500 and 2,000 mg kg^{-1} mice for three days and found no genotoxic effect in blood and liver samples of these animals by the comet test. However, the aqueous extract showed antigenotoxic potential (oxidative damage induced by hydrogen peroxide) in blood samples from these mice by the comet test, which may be due to the antioxidant potential demonstrated *in vivo by the* aqueous extract of this plant, which may be responsible for protecting DNA molecules against oxidative damage induced by free radicals.

Regarding the antioxidant potential of this plant, Pádua et al. (2010) reported a good antioxidant potential *in vitro and in vivo* for the hydroethanolic extract of *B. trimera in the* neutrophils of Fisher rats. Mendes, Tabach and Carlini (2007), on the other hand, reported that the hydroalcoholic extract of *B. trimera* has a moderate antioxidant capacity and is not very efficient as an anti-ulcer in rats (stress ulcer induction). However, Dias et al. (2009), analyzing different extracts of *B. trimera,* observed that some extracts showed anti-ulcer activity and were able to reduce lesions in *Wistar* rats (ulcer induction by hydrochloric acid in ethanol).

With regard to the treatment of obesity, Souza et al. (2011) found high inhibitory activity of methanolic extracts of *B. trimera* on some digestive enzymes (lipase and amylases), showing that the use of these extracts may be useful to help treat this disease.

In Brazil, *B. trimera* is among the most commercialized medicinal plants, and more detailed studies on its chemical constituents and how it acts in the treatment of obesity could contribute to a safer use of this plant.

2.7.2 *Garcinia cambogia*

Garcinia cambogia Desr. is a small tree belonging to the Gutiferaceae family, native to the forests of Cambodia, southern Africa and Polynesia, widely cultivated in India and Far Eastern countries. The most commonly used parts of the plant are the dried bark and the pulp of the fruit, the fruit being the part that contains the component capable of promoting weight loss (SANTOS et al., 2007).

The hydroxycitric acid (HCA) found in the skin of the fruit is the substance responsible for the fat loss and satiety properties of *G. cambogia* (SANTOS et al., 2007). The following actions are attributed to HCA for promoting weight loss:

a) Lipid blocker - when ingested in excess, carbohydrates are metabolized and stored in the form of lipids, in which the enzyme citrate lyase is involved. AHC binds to this enzyme and inhibits it, preventing the storage of lipids.

b) Appetite suppressant - appetite control is based on increased glycogen synthesis by AHC, which, by blocking citrate lyase, directs calories that are not stored in the form of lipids to the formation of glycogen, so when glycogen reserves are high, sugar receptors in the liver are stimulated and send a satiety signal to the brain, without stimulating the central nervous system. Another mode of action is related to its ability to stimulate the release of serotonin, a neurotransmitter directly involved in appetite control.

Studies carried out with *G. cambogia* extracts have *shown* a reduction in visceral adiposity in rats treated with a high-fat diet (KIM et al., 2008), providing leptin-like activity in reducing blood insulin levels and improving glucose metabolism (HAYAMIZU et al., 2003a), and in humans a reduction in abdominal fat typical of obesity, suggesting its use in preventing and reducing the accumulation of visceral fat (HAYAMIZU et al., 2003b). Under conditions of diets rich in carbohydrates, isolated AHC was effective in reducing the synthesis of new lipids in humans (KOVACS; WESTERTERP-PLANTENGA, 2006).

As for its toxicity, Saito et al. (2005) observed that diets containing *G. cambogia* (with hydroxycitric acid contents of more than 10.10 g kg^{-1} of feed) administered to obese Zucker rats proved to be highly toxic to their testes and caused a decrease in spermatogenesis after 13 weeks of treatment. In *Wistar* rats, ethanolic extracts of *G. cambogia* administered for six weeks showed dose-dependent toxic effects on the reproductive system, with distortions identified in the cells of the germinal epithelium (KAYODE et al., 2007).

For humans, no adverse effects were observed from *G. cambogia* extracts administered in tablet form, containing 1000 mg of hydroxycitric acid per day^{-1} , for 16 weeks (HAYAMIZU et al., 2003b), and for 12 weeks, no changes were observed in serum levels of the sex hormones testosterone, estrone and oestradiol (HAYAMIZU et al., 2008).

Although AHC has a relatively low toxicity, the broad spectrum of its actions and the scarcity of information on its long-term effects mean that more detailed studies are needed on the possible interference with certain physiological parameters.

2.7.3 *Tournefortia paniculata*

Tournefortia paniculata Cham., traditionally known as marmalade, is a shrub belonging to the Boraginaceae family. Its origin is attributed to the Mediterranean regions and the United States of America, occurring widely in tropical and subtropical regions (MORAES; SOUSA, 2007). In Brazil, its leaves are used in folk medicine as a diuretic and anti-inflammatory for the urinary tract and in cases of kidney lithiasis (BERTOLUCCI et al., 2000). Few studies have been carried out on its chemical constituents, thus requiring extensive chemical and functional characterization.

2.7.4 *Simaba ferruginea*

Simaba ferruginea St. Hil, popularly known as calunga, calunga ferruginea, or fel da terra, with the scientific synonym *Quassia ferrugiena* D. Dietr. Dietr, is a tree native to the dry cerrado of central Brazil, belonging to the Simaroubaceae family. The bark of both the trunk and the rhizome contains bitter-tasting substances, which are used in folk medicine to treat fever, wound healing, rheumatic pain, gastric ulcers, obesity and other ailments, mainly in the form of teas (LORENZI; MATOS, 2002; MARCELO, 2002). Although several plants belonging to the Simaroubaceae family have been used in traditional medicine to treat various diseases, only a few chemical and pharmacological products have been reported on the *Simaba* genus (MUHAMMAD et al., 2004; TOMA et al., 2002), which highlights the need for studies on its chemical composition and pharmacological activity to prove its efficacy, which has so far only been reported in folk medicine.

2.8 Digestive enzyme inhibitors

Digestive enzyme inhibitors are chemical substances (mostly proteins) found in plant tissues, such as seeds, roots and others, and in animals, such as egg whites. These substances, when ingested, inhibit the action of enzymes such as amylases, lipases and trypsin, which are very important for the normal metabolism of the human body, thus preventing the absorption of carbohydrates, lipids and proteins, which can lead to weight loss and can be an important tool to help treat obesity (GENOVESE; LAJOLO, 2001).

The presence of these inhibitors in natural products has been widely studied in the search for drugs based on the mechanism of enzyme inhibition that causes beneficial changes in metabolism, presenting itself as an excellent alternative for the safe and effective development of anti-obesity drugs (BHUTANI; BIRARI; KAPATI, 2007).

Many studies have demonstrated the efficacy, importance and potential of using inhibitors of these enzymes in the treatment of obesity and associated comorbidities and reinforce the need to search for new sources of these inhibitors: amylases (OBIRO; ZHANG; JIANG, 2008; UDANI et al., 2009), glycosidases (KWON; APOSTOLIDIS; SHETTY, 2006) and lipases (SHARMA; SHARMA; SEO, 2005; SOUZA et al., 2011). Thus, digestive enzyme inhibitors that help limit the intestinal absorption of carbohydrates and fats in the initial phase may prove useful as aids in the treatment of obesity.

2.8.1 Amylases

Amylases are some of the enzymes produced by animals, plants and microorganisms that take part in the hydrolysis of carbohydrates to enable them to be used as a source of energy. For this energy to be used by living beings, carbohydrates must be hydrolyzed into monomers and absorbed. In this way, they can participate in intracellular catabolic processes, releasing the energy they store (WANG et al., 2008).

The main form of carbohydrate storage in plants, including those used in human food, is starch, a nutritional reserve complex made up of two types of glucose polymers, amylose and amylopectin, with amylose being an unbranched glucose polymer linked by a-1,4 glycosidic bonds and amylopectin a branched polymer with a main glucose chain linked by a-1,4 glycosidic bonds interconnected by a-1,6 bonds (SORENSEN et al., 2004).

Therefore, due to the complexity of starch, many enzymes are required for its complete degradation and consequent absorption and use as an energy source. These are divided into groups according to their mechanism of action: endo-amylases, exo-amylases and deamylating enzymes.

Endo-amylases, known as a-amylases (1,4-glucan-4-glucanohydrolases), classified by the acronym E.C 3.2.1.1, cleave the a-1,4 glycosidic bonds present in the inner part of starch, converting it into simple sugars so that they can be completely digested by the body. The hydrolysis carried out by this enzyme takes place in several stages, starting with the binding of the substrate to the enzyme, followed by the separation of the polymer and a hydrolytic process to release several smaller molecules (ANTUNES, 2008). The products resulting from digestion by a-amylase are a mixture of maltose, maltotriose and branched glucose units.

The inhibition of a-amylase can have a wide range of applications, such as in the adjuvant treatment of obesity and type 2 diabetes, since its inhibition can hinder the digestion of carbohydrates, thus reducing the absorption and consequently the availability of these sugars, promoting or supporting weight loss (CHOKSHI, 2007). Another field of application has been in agriculture to control insects and pests.

Several types of organic compounds are known for their amylase inhibitory activity. Non-protein inhibitors are generally low molecular weight molecules, such as phenolic compounds and sugars, while protein inhibitors are classified according to their structure, with one group found in fungi and the other six coming from higher plants (SILVA, 2008). Amylase inhibitors have already been identified in various plant species, and the discovery of new plants with inhibitory potential could become a strong ally in helping to treat obesity and its complications.

Another group of enzymes used for starch degradation are exo-amylases, represented by a-glucosidase, which cleaves the external a- 1,4 glycosidic bonds of amylose, amylopectin and related polysaccharides (OTA et al., 2009).

Inhibitors of a-glucosidases are also agents of great therapeutic interest since, by inhibiting the degradation of sugars for absorption by the body, they are active against obesity (KANDRA et al., 2005). Other applications of these inhibitors include the treatment of viral infections, tumors, osteoarthritis and other diseases.

Some drugs such as arcabose and miglitol are used as inhibitors of a-glucosidases and are used in the treatment of obesity, but apart from these, there are few other therapeutic options, justifying the need for studies to obtain new sources of inhibitors of this enzyme.

2.8.2 Lipases

Lipases are described as triacylglycerol lipases (E.C.3.1.1.3), which act on the ester bonds present in triglycerides, releasing fatty acids and glycerol, constituting a special class of carboxylesterases. They mainly hydrolyze long-chain triglycerides, i.e. those with an acyl chain of more than 10 carbon atoms. These enzymes are of great importance, not only because they act on water-insoluble substrates, but also because they catalyze different reactions, such as hydrolysis, esterification, alcoholysis and aminolysis (MUKHERJEE, 2003). This diversity of properties allows lipases to be used in different fields of application.

Pancreatic lipase is the main lipolytic enzyme synthesized by the pancreas and plays an important role in the efficient digestion of triglycerides. It acts by removing fatty acids, preferably in positions 1 and 3 of the triglycerides ingested in the diet, releasing 2-monoacylglycerol and long-chain saturated and polyunsaturated fatty acids as lipolytic products, being responsible for the hydrolysis of 50 to 70% of the total fats ingested in the diet (MUKHERJEE, 2003; SHI; BURN,

2004).

Pancreatic lipase inhibition is a very promising option to help treat obesity, as it can limit the nutritional impact of lipid absorption, and is one of the most studied mechanisms to determine the potential effectiveness of natural products as anti-obesity agents. Anti-lipase agents act by reducing or blocking the digestion of lipids, preventing their absorption from the gastrointestinal tract and therefore mimicking the effect of reduced food intake.

One of the main drugs used in the treatment of obesity, orlistat, acts by inhibiting lipases, and its success has stimulated a lot of research to identify new inhibitors of these enzymes. Among this research is the search for natural products with inhibitory action on lipase, which has been the subject of many studies, mainly due to the discovery in plants of inhibitory molecules belonging to various classes of compounds such as saponins, lectins and some phenolic compounds (BIRARI; BHUTANI, 2007; SHARMA; SHARMA; SEO, 2005; SOUZA et al., 2011).

2.8.3 Trypsin

Trypsin (E.C.3.2.21.4) is a digestive enzyme produced by the pancreas in the form of its enzymatically inactive zymogen (trypsinogen) and taken to the intestine, where through the enzymatic action of enterokinase and/or trypsin it becomes active and digests proteins, releasing peptides (BARCELOS, 2004). It belongs to the class of endopeptidases, i.e. it is a protein that hydrolyzes peptide bonds far from the C and N-terminal ends. In addition to intestinal digestion, it is involved in the processes of cell lysis of invading organisms, blood coagulation, fertilization, among others.

Trypsin is susceptible, to varying degrees, to some natural protease inhibitors, which can be protein or non-protein in nature (SCHACHTER, 1980). Blocking the action of this enzyme results in an excessive increase in the plasma concentration of cholecystokinin and, as a result, the pancreas is continually stimulated to release more enzyme, causing pancreatic hypertrophy (SILVA; SILVA, 2000). Although pancreatic hypertrophy is frequently reported in laboratory animals and is supposedly triggered by the presence of protease inhibitors, there is no evidence of harmful effects in humans, even in the case of legumes consumed raw, such as peanuts, which have been shown to have trypsin inhibitors, and there have been no reports of any harmful effects on health (SGARBIERI, 1987).

Some studies show that inhibiting the digestion of proteins can be beneficial to health, as it reduces the total caloric value of a meal and slows down its digestion, leading to a feeling of satiety during the passage of the food bolus through the gastrointestinal tract, which can contribute to weight loss. However, as the amino acids obtained from protein digestion are essential substances for the body, trypsin inhibition can also cause problems for the body and is considered an anti-nutritional effect (MCDOUGALL et al., 2005).

2.9 Free radicals and antioxidants

Free radicals (FR) are chemical species capable of independent existence, containing one or more unpaired electrons, with a very short half-life. As such, they are highly reactive and capable of attacking any biomolecule, characterizing them as destructive molecules (BIANCHI; ANTUNES, 1999).

The most important radicals present in living organisms are hydroxyl radicals (HO*), superoxide anion ($O2^{-}$ *), peroxyl (ROO'), alkoxyl (RO') and nitric oxide ('NO). There are other reactive species in organisms that are not free radicals but can induce radical reactions, such as singlet oxygen (^{1} O2), hydrogen peroxide (H2O2) and hypochlorous acid (HOCl) (YILDIRIM; MAVI; KARA, 2001).

These radicals can be generated by endogenous or exogenous sources. Endogenous sources originate from biological processes that occur normally in the body, such as the reduction of flavins and thiols; the result of the activity of oxidases, cyclooxygenases, lipoxygenases, dehydrogenases and peroxidases; the presence of transition metals inside the cell and electron transport systems. This generation of RL involves various cell organelles, such as mitochondria, lysosomes, peroxisomes, nucleus, endoplasmic reticulum and membranes. Exogenous sources include smoke, air pollution, organic solvents, anesthetics, pesticides and radiation (SOARES, 2002; YILDIRIM; MAVI; KARA,

2001).

Free radicals play an important role in the body, acting against infectious organisms, helping to maintain the immune system, among other things (HALLIWELL, 2011). Generally, the metabolism balances their presence in the body, but the problem is that when they form in excess, under certain abnormal conditions, a situation is created which is called oxidative stress (BARREIROS; DAVI; DAVI, 2006). It is now known that oxidative stress is associated with various processes of cellular toxicity, such as damage to the structure of proteins, lipid peroxidation in membranes, alteration of DNA and inactivation of enzymes (BIANCHI; ANTUNES, 1999).

In this way, they are involved in various pathologies, such as diseases of the gastrointestinal tract, such as gastric and intestinal ulcers; diseases related to the nervous system, such as Down's syndrome and Alzheimer's disease; cardiovascular diseases, such as atherosclerosis and cancer. All the damage caused to the body can result in premature ageing, which can be the cause or aggravation of these diseases (HALLIWELL; GUTTERIDGE; CROSS, 1992). In addition to being the cause of various diseases, LRs also cause damage to food, affecting its nutritional quality, color, aroma, taste, texture, and are responsible for the formation of potentially toxic and anti-nutritional compounds for the body.

With the discovery of the deleterious effects caused by RL in the human body and in food, great interest has arisen in the study of substances known as antioxidants. Antioxidants are substances capable of slowing down or inhibiting the oxidation of oxidizable substrates and their main function is to protect cellular constituents and maintain the cellular redox state (HALLIWELL, 2001). They prevent the deleterious effects of oxidation by inhibiting the onset of lipoperoxidation, sequestering RL and/or chelating metal ions; they protect aerobic organisms from oxidative stress (RODRIGUES et al., 2003). Although antioxidants donate electrons to certain substrates, they are not transformed into LR, since they are stable in both forms.

The body has an enzymatic system (endogenous) to combat RLs, consisting of several enzymes, including superoxide dismutase (SOD), catalase (CAT) and glutathione peroxidase (GPx); the first to act prevents the accumulation of hydrogen peroxide and radicals such as superoxide (HALLIWELL, 2011).

SOD plays a fundamental role in the body's defense against reactive oxygen species, as it acts to remove the superoxide radical. CAT breaks down excess hydrogen peroxide, a toxic product of the body, into water and molecular oxygen, which is also done by the GPx enzyme, but it only acts in this reaction when hydrogen peroxide is present in low concentrations (BARREIROS; DAVI; DAVI, 2006; BIANCHI; ANTUNES, 1999; ROVER JÚNIOR; HOEHR; VELLASCO, 2001).

Despite the body's high defense capacity, endogenous antioxidant defenses are not infallible and do not fully protect cellular components. It is therefore well established that antioxidants obtained from exogenous sources are indispensable for adequate defense against LDs, thus playing an important role in maintaining health.

Research has shown that many vegetables, fruits and medicinal plants contain compounds with antioxidant action, including vitamins C and E, phenolic compounds, carotenoids and some minerals (DJERIDANE et al., 2006). Many of these phytochemicals have an antioxidant capacity that is associated with a lower occurrence of some human diseases.

Vitamin C acts as an antioxidant because ascorbate can donate hydrogen atoms to reduce another compound. It acts in the aqueous phase as an excellent antioxidant in LRs, but is unable to act in lipophilic compartments to inhibit lipid peroxidation.

In phenolic compounds, the antioxidant potential is attributed to the redox properties of the hydroxyl groups and their relationship with different parts of the chemical structure which can be determined by five factors: reactivity as a hydrogen and electron donating agent, stability of the flavanoyl radical formed, reactivity against other antioxidants, ability to chelate transition metals and solubility and interaction with membranes (BARREIROS; DAVI; DAVI, 2006). These compounds can act in both the initiation and propagation phases of the oxidative process and the intermediate products formed by their action are relatively stable due to the resonance of their aromatic ring (RAMALHO; JORGE, 2006; SOARES, 2002).

Carotenoids act *in vivo* to combat LDs, acting as singlet and triplet oxygen deactivators or as peroxyl radical scavengers. They are able to interrupt the reactions of RLs that can oxidize unsaturated lipids and protect DNA from attack by these radicals. The ability of carotenoids to exterminate LRs is due to their structure containing alternating double bonds, which have the ability to accommodate unpaired charges or electrons.

Carotenoids act as lipophilic antioxidants and, together with vitamin C and phenolic compounds (hydrophilic antioxidants), form a strong defense against LRs, acting in different cellular compartments.

Thus, these compounds show potential for preventing and combating RL and the various diseases caused by it.

3 FINAL CONSIDERATIONS

The prevalence of obesity is increasing at an alarming rate worldwide and is becoming the biggest health problem in modern society, highlighting the need for adjuvants to help treat it. Medicinal plants are rich sources of bioactive compounds with potential for therapeutic use, which could be a viable alternative for the development of effective and safe drugs to help treat obesity.

This study found that the plants studied, *Aloe vera* (L.) Burm. (aloe), *Simaba ferruginea* St. Hil. (calunga), *Baccharis trimera* (Less.) DC (carqueja), *Garcinia cambogia* Desr. and *Tournefortia paniculata* Cham. (marmalade), have a great diversity of phytochemicals with potential for use in pharmacological preparations with possible health benefits. Among the plants, *T. paniculata has* the highest levels of phytochemicals (phenolic compounds, saponins and vitamin C), high inhibition of the enzymes a-amylase and a-glucosidase and antioxidant activity *in in vitro* tests.

The evaluation of the therapeutic and toxicological effects of *T. paniculata* revealed that the leaves of this plant have anti-obesity and antioxidant potential, as they cause a reduction in glucose levels, triacylglycerols and liver fat (flour and aqueous extract) and food consumption and lipid peroxidation (aqueous extract), as well as not causing genotoxic effects under the conditions evaluated. However, a more detailed study of the medicinal and toxicological potential, as well as the characterization of the phytochemicals present in *T. paniculata is* still necessary to better understand its mechanisms in the treatment of obesity, as well as other possible therapeutic applications.

CHAPTER II

ARTICLES

ARTICLE 1

CHEMICAL COMPOSITION OF MEDICINAL PLANTS USED AS AIDS IN THE TREATMENT OF OBESITY

Anderson Assaid Simão; Angelita Duarte Corrêa

Submitted to Bioscience Journal.
Article written in accordance with the journal's standards. This article is a preliminary version, as the journal's editorial board may suggest changes to adapt it to its own style.

ABSTRACT: Medicinal plants are rich sources of bioactive compounds with potential therapeutic use, and knowledge of the composition of the compounds present in their extracts is essential to ensure scientific proof of the efficiency and safety of their use. The aim of this study was to carry out a phytochemical screening, determine the percentage and mineral composition, as well as the content of some bioactive compounds in the medicinal plants *Aloe vera* (L.) Burm. (aloe), *Simaba ferruginea* St. Hil. (calunga), *Baccharis trimera* (Less.) DC (carqueja), *Garcinia cambogia* Desr. and *Tournefortia paniculata* Cham. (marmalade), with the aim of finding substances of pharmacological interest in order to help treat obesity and other diseases. Important substances with therapeutic potential, especially phenolic compounds, saponins and dietary fiber, were found in all the plants, as well as calcium in *G. cambogia* and *S. ferruginea*. It was possible to conclude that the plants studied have a great diversity of phytochemicals with the potential to be used in pharmaceutical preparations with possible health benefits. However, recommending the use of these plants in relation to possible risks and benefits for human health is premature, and further toxicity, efficiency and safety studies are needed, particularly in relation to the saponins found in all the plants and the high levels of phenolic compounds in *T. paniculata* (36.19 g 100 g^{-1} dry matter).
KEYWORDS: Phytochemical screening. Percentage composition. Minerals. Bioactive compounds. Medicinal plants. Obesity.

INTRODUCTION

The prevalence of obesity has increased progressively and is considered a major public health problem in developed countries and a global epidemic by the World Health Organization (WORLD HEALTH ORGANIZATION - WHO, 2010). Its treatment is essential, as it is associated with various diseases, such as diabetes, some types of cancer, cardiovascular diseases, among others (GUH et al., 2009).

Despite being one of the oldest diseases known to man, pharmacological options for treating obesity are still limited and have many side effects (MAHAN; SCOTT-STUMP, 2008). In view of the above, the use of medicinal plants has been widely explored, both by the population, due to their easy access, low cost, lack of need for prescriptions and belief in the absence of toxic effects, and by the pharmaceutical industry, which sees these plants as a viable alternative for the future development of drugs that induce weight reduction in an effective and safe way (MAYER et al., 2009). Studies show that various natural products, including extracts and compounds isolated from plants, are used to reduce body weight and prevent obesity (SOUZA et al., 2011; SIMAO et al., 2012).

Several medicinal plants, such as *Aloe vera* (L.) Burm. (aloe), *Simaba ferruginea* St. Hil. (calunga), *Baccharis trimera* (Less.) DC (carqueja), *Garcinia cambogia* Desr. and *Tournefortia paniculata* Cham. (marmalade), are used to treat obesity (SOUZA et al., 2011; SIMÂO et al.; 2012; 2013). However, most of these plants have no scientific proof of their efficacy and safety in the treatment of this disease, and their use is based solely on popular beliefs, highlighting the need for scientific studies to elucidate the chemical constituents of these plants, in order to aid in the pharmacological/toxic understanding of their use in the treatment of obesity and other diseases.

The aim of this study was to carry out phytochemical screening, determine the chemical and percentage compositions, as well as the content of some bioactive compounds of the medicinal plants *A. vera*, *B. trimera*, *S. ferruginea*, *G. cambogia* and *T. paniculata*, used as aids in the treatment of obesity, in order to find substances of pharmacological interest and guarantee the use of these plants by the population in industrialized and herbal formulations.

MATERIAL AND METHODS

Sample collection and preparation

The leaves of *B. trimera* and *T. paniculata*, as well as the stem bark of *S. ferruginea,* were purchased at the municipal market in Belo Horizonte, Minas Gerais, in January 2011 and transported to the laboratory. The leaves of *B. trimera* and *T. paniculata* were washed with running distilled water and then placed together with the stem bark obtained from *S. ferruginea* in a forced air circulation oven to dry for 48 hours at a temperature of ±35°C. After drying, the leaves and bark were ground in a Wiley mill, and the flours were stored in hermetically sealed jars until the analyses were carried out. The commercial powder of *A. vera*, obtained by freeze-drying the plant's mucilage, and *G. cambogia*, obtained by spray-drying, were purchased from FLORIEN, a distributor of pharmaceutical raw materials.

Analysis

Phytochemical screening

The medicinal plant flours (MPF) were subjected to phytochemical screening using specific reagents for each chemical group, using chemical reactions that result in the development of a color and/or precipitate characteristic of each class of substance (MATOS, 1997). Organic acids, reducing sugars, alkaloids, anthraquinones, azulenes, carotenoids, catechins, depsides and depsidones, coumarin derivatives, steroids and triterpenoids, flavonoids, cardiotonic glycosides, sesquiterpene lactones and other lactones, polysaccharides, proteins and amino acids, saponins and tannins were analyzed.

Percentage composition

Moisture content was determined in an oven at 105°C until constant weight. The ether extract was determined using a continuous Soxhlet extractor. Crude protein was measured by the Kjeldahl method, using a conversion factor of 6.25 (N x 6.25). Ash and fixed mineral residue were obtained from a defined quantity of samples by incineration (550°C) in a muffle furnace, thus determining the percentage of residue. Total, soluble and insoluble dietary fiber was determined using the enzymatic method. The non-nitrogenous extract was determined by the difference between 100 and the sum, in dry matter, of the ether extract, protein, ash and total dietary fiber. These percentage composition analyses were carried out according to the methodology described by the Association of Official Analytical Chemists - AOAC (2005).

Mineral composition

To quantify the minerals (Fe, Zn, Mn, Cu, Ca, Mg, P, K and S), the samples were subjected to nitroperchloric digestion in temperature-controlled digestion blocks. P and S were determined by colorimetry, K by flame photometry and Ca, Mg, Cu, Mn, Zn and Fe by atomic absorption spectrophotometry. For all the analyses, the procedures described by Malavolta et al. (1997) were used.

Phenolic compounds

The phenolic compounds were extracted using 50% methanol under reflux for three consecutive times at 80°C, and the extracts were collected, evaporated to 25 mL and dosed for phenolic compounds using the Folin-Denis reagent and tannic acid as a standard (AOAC, 2005).

Oxalic acid

To determine oxalic acid, the method developed by Loures and Jokl (1990) was used; according to this method, oxalic acid was extracted hot with hydrochloric acid, precipitated and quantified by titrating calcium oxalate with potassium permanganate.

Nitrate

The nitrate was extracted with distilled water at 45°C. In the dosage, a complex is formed by the nitration of salicylic acid under highly acidic conditions, and the reading at 410 nm is taken in basic solutions (pH greater than 12). The absorbance of the material was directly proportional to the amount of nitrate present, with no interference from ammonium, nitrite or chlorine ions. Potassium nitrate was used as a standard (CATALDO et al., 1975).

Trypsin inhibitor

The trypsin inhibitors were extracted with 0.1 mol L^{-1} NaOH under magnetic stirring. After centrifugation at 10,000 x g for 60 minutes, an aliquot of the supernatant was used in the enzymatic assay (KAKADE et al., 1974). Trypsin activity was determined according to the methodology proposed by Erlanger et al. (1961), in which 200 |iL of the plant extracts and 200 til. of enzyme were incubated in a water bath at 37°C for four periods of time after the addition of 800 LI L. of the substrate benzoyl-DL-arginine-p-nitroanilide (BApNA) prepared in TRIS buffer (tris(hydroxymethyl)aminomethane) at 0.05 mol L^{-1} , pH 8.2 plus 20 mmol L^{-1} CaCl2. The reaction was stopped by adding 200 ml of 30% acetic acid and the product was read in a spectrophotometer at 410 nm.

The inhibition of trypsin was obtained by determining the slopes of the lines (absorbance vs. time) of the activity tests of the control enzyme (without plant extract) and the enzyme + inhibitor (with plant extract). The slope of the line is due to the rate of product formation per minute of reaction and the presence of the inhibitor causes a reduction in the slope. The absorbance values were converted into nmol using the molar extinction coefficient of BApNA determined by Erlanger et al. (1961).

Saponins

The saponins were extracted with ethanol under agitation at room temperature for 60 minutes. The total saponin content was determined by reacting the saponins with anisaldehyde, using digitonin as a standard (BACCOU et al., 1977).

Statistical analysis

The data are the mean of three replicates ± standard deviation and were statistically evaluated by analysis of variance, with the means being compared using the Scott Knott test ($P<0.05$) with the aid of the R software (R Development Core Team, 2011).

RESULTS AND DISCUSSION

Phytochemical screening is characterized by the identification of the chemical compounds present in the plant material. The results of the phytochemical screening on the FMPs used as aids in the treatment of obesity are shown in Table 1.

The results indicate the presence of different metabolic groups of pharmacological interest in the plants studied, such as tannins (*G. cambogia* and *T. paniculata*), which prevent lipid peroxidation, nucleotide degradation and accelerate the healing process (PIETTA et al., 2000; MONTEIRO et al.; 2005); depsides and depsidones (*B. trimera*, *S. ferruginea* and *T. paniculata*), which have antioxidant properties (HIDALGO et al. , 1994), antiviral properties (NEAMATI et al., 1997), antitumor properties, analgesic and

Table 1. Phytochemical screening of medicinal plants.

Components	*Aloe vera*	*Baccharis trimera*	*Simaba ferruginia*	*Garcinia cambogia*	*Tounerfotia paniculata*
Organic acids	(-)	(-)	(-)	(-)	(+)
Reduce sugars	(+)	(+)	(+)	(+)	(+)

Alkaloids	(-)	(-)	(-)	(-)	(-)
Anthraquinones	(-)	(-)	(-)	(-)	(+)
Azulenes	(+)	(+)	(+)	(-)	(-)
Carotenoids	(-)	(+)	(-)	(-)	(+)
Catechins Depsides	(-)	(-)	(-)	(+)	(+)
and coumarin depsidones	(-)	(+)	(+)	(-)	(+)
derivatives Steroids and	(-)	(-)	(-)	(-)	(-)
triterpenoids	(-)	(+)	(+)	(-)	(+)
Flavonoids	(-)	(+)	(+)	(-)	(+)

Cardiac glycosides Sesquiterpene	(-)	(-)	(-)	(-)	(-)
lactones and other lactones	(-)	(-)	(-)	(-)	(+)
Polysaccharides Proteins and amino acids	(+)	(+)	(-)	(-)	(-)
acids	(+)	(+)	(+)	(+)	(+)
Saponins	(+)	(+)	(+)	(+)	(+)
Tannins	(-)	(-)	(-)	(+)	(+)

The (+) sign indicates the presence and (-) indicates the absence of the metabolite.
antipyretic properties (OKUYAMA et al., 1995); carotenoids (*B. trimera* and *T. paniculata*), with properties associated with various physiological actions, may be precursors of vitamin A, and are implicated in protective mechanisms against oxidative damage to cells, cardiovascular diseases, cataracts and macular degeneration, as well as contributing to improving the immune system (SIMOES et al, 2007); triterpenoids (*B. trimera, S. ferruginea and T. paniculata), with medicinal properties and great potential for biological activities, such as anti-inflammatory and anti-inflammatory properties (OKUYAMA* et al, 1995). *trimera*, *S. ferruginea* and *T. paniculata*), with medicinal properties and great potential in biological activities, such as anti-inflammatory, bacterial, fungicidal, antiviral, analgesic, cardiovascular, antitumor, etc. (SIMOES et al., 2007), among other metabolic groups, 2007), among other groups of metabolites.

Alkaloids and cardiac glycosides were not detected in any of the plants analyzed. The preliminary phytochemical screening provides an overview of the chemical groups present in the plants, but further studies are needed to determine the concentration and characterization of these substances.

The results of the phytochemical screening are in line with other studies with these plants, which have also shown the presence and absence of these metabolites, such as the study by Vasquez et al. (1996), with *A. vera* gel and its extracts, which recorded the presence of saponins and the absence of tannins, flavonoids and alkaloids; those by Rodrigues et al. (2009), with *B. trimera, who observed the* presence of flavonoids, saponins and the absence of alkaloids, anthraquinones, coumarins and cardiac glycosides; those of Moraes & Souza (2007), with the leaves of *T. paniculata, who reported the presence of tannins,* flavonoids and alkaloids. *paniculata leaves*, which reported the presence of flavonoids, tannins and no alkaloids, and those of Subhashini et al. (2011), with *G. cambogia,* which found saponins, tannins, sugars, proteins and no flavonoids, but the alkaloid test result was positive. In the literature consulted, no studies were found on the phytochemical screening of S. *ferruginia.*

The results of the chemical composition of the FMPs are shown in Table 2. With regard to ether extract, the levels found were low, with *T. paniculata* having the highest content (3.88 g 100 g^{-1} dry matter - DM). Ether extract was not detected in *A. vera* and *G. cambogia.* The crude protein content was also relatively low, and the ash content was very high in *G. cambogia* (34.18 g 100 g^{-1} DM).

The highest dietary fiber (DF) content was found in *B. trimera* (66.26 g 100 g^{-1} DM), and the lowest in *A. vera* (10.97 g 100 g^{-1} DM).

Table 2. Percentage composition, in g 100 g^{-1} dry matter, of medicinal plants.

Components	*Aloe vera*	*Simaba ferruginea*	*Baccharis trimera*	*Garcinia cambogia*	*Toumefortia paniculata*
Ether extract	ND[2]	1.97 ± 0.11^{c}	2.49 ± 0.20^{b}	ND	3.88 ± 0.21^{a}
Crude protein	1.54 ± 0.11^{d}	8.96 ± 0.11^{b}	7.15 ± 0.22^{c}	1.78 ± 0.25^{d}	10.78 ± 0.35^{a}
Ashes	3.30 ± 0.08^{e}	8.53 ± 0.04^{b}	6.18 ± 0.05^{c}	34.18 ± 0.24^{a}	3.61 ± 0.01^{d}
Insoluble fiber	5.20 ± 0.18^{d}	45.20 ± 0.43^{c}	64.67 ± 0.50^{a}	3.09 ± 0.18^{d}	46.71 ± 0.80^{b}
Soluble fiber	5.77 ± 0.33^{b}	3.74 ± 0.22^{c}	1.60 ± 0.07^{d}	30.11 ± 1.19^{a}	1.16 ± 0.05^{d}
Total fiber	10.97 ± 0.42^{d}	48.94 ± 0.62^{b}	66.26 ± 0.53^{a}	33.08 ± 1.02^{c}	47.87 ± 0.86^{b}
NNE[1]	84.18 ± 0.51^{a}	31.60 ± 0.64^{c}	17.93 ± 0.60^{d}	30.96 ± 1.17^{c}	33.86 ± 0.11^{b}

Data are the mean of three replicates ± standard deviation. The same letters in the rows do not differ by the Scott-Knott test (P <0.05). 'NNE: Non-nitrogenous extract. [2]ND: Not detected. Moisture content of medicinal plant flours, in g 100 g'[1] : *Aloe* = 8.53; *Simaba* = *8.*42; *Baccharis* = *8.*56; *G. cambogia* = 3.94; *Tourneforticr.* 9.90.

The high content of soluble FD shown by *G. cambogia* (30.11 g 100 g^{-1} DM), as well as insoluble FD found in *B. trimera* and *T. paniculata*, 64.67 and 46.71 g 100 g^{-1} DM, respectively, should also be mentioned.

Epidemiological studies suggest that dietary fiber is capable of preventing obesity and weight gain, as well as helping to reduce the risk of developing diabetes and cardiovascular diseases, among others (LIU et al., 2003).

In general, soluble FD helps in the treatment of obesity, as it delays gastric emptying, glucose absorption and reduces cholesterol in the blood serum (RIQUE et al., 2002; MELO; LAAKSANEM, 2009). On the other hand, insoluble DM speeds up intestinal transit and increases stool weight (RIQUE et al., 2002). Thus, the presence of these fibers in the composition of the plants analyzed can be characterized as a strong ally in the treatment of obesity.

The non-nitrogenous extract or glycidic fraction consists mainly of sugars. Thus, the highest content was found in *A. vera* (84.18 g 100 g^{-1} DM), followed by *T. paniculata*, *S. ferruginea* and *G. cambogia.*

In addition to the influence exerted by macronutrients on the development of obesity, micronutrients, especially minerals, have received a lot of attention due to their influence on body weight control.

Table 3 shows the mineral content of the FMPs used as aids in the treatment of obesity. It is important to highlight the high contents, in mg 100 g^{-1} DM, of potassium (1,009.44) and manganese (47.07) in *A. vera,* potassium (1,383.13), calcium (2,660.70) and iron (55.67) in *S. ferruginea, potassium (*2,336.69) and iron (39.98) in *B. trimera* and calcium (7,273.23), sulphur (999.38) and iron (73.69) in *G. cambogia.*

Minerals play important roles in the human body, one of which is the regulation of metabolism. The absence of some minerals can cause metabolic problems, such as slowing down the metabolism, which can lead to weight gain. In addition, some minerals can participate in the digestion of carbohydrates, fats and proteins, and can act as aids in weight reduction.

The high levels of calcium found in *G. cambogia* and *S. ferruginea* can be extremely effective in treating obesity, since calcium intake is involved in regulating body weight (ST-ONGE, 2005).

Table 3. Mineral composition, in g 100 g'[1] dry matter, of medicinal plants.

Mineral	*Aloe vera*	*Simaba ferruginea*	*Baccharis trimera*	*Garcinia cambogia*	*Tournefortia paniculata*
P	44 40±0.00^{c}	83.84±6.31^{a}	83.84±6.31^{a}	76.44±0.00^{b}	44 40±0.00^{c}
K	1,009.44±6.31^{c}	1,383.13±27.48^{b}	2,336.69±49.31^{a}	742.59±69.31^{d}	495.75±12.82^{e}
Ca	845.45±6.31^{c}	2,660.70±16.68^{b}	612.42±0.00^{d}	7,273.23±33.46^{a}	381.06±23.10^{e}

Mg	258.74±6.31^{b}	163.79±0.00^{c}	142.17±0.00^{d}	419.88±6.01^{a}	147.98±6.41^{d}
s	258.74±27.51^{e}	731.60±10.92^{b}	630.65±12.63^{c}	999.38±63.32^{a}	527.19±23.54^{d}
Cu	13.12±0.21^{a}	3.10±0.32^{c}	7.81±0.76^{b}	3.38±0.38^{c}	2.04±0.27^{d}
Mn	47.07±0.49^{a}	5.24±0.11^{e}	13.47±0.27^{c}	18.75±0.11^{b}	7.00±0.17^{d}
Zn	1.57±0.23^{d}	2.20±0.16^{c}	3.84±0.00^{b}	6.51±0.15^{a}	1.40±0.05^{d}
Fe	2.58±0.38^{e}	55.67±0.16^{b}	39.38±1.44^{c}	73.69±0.65^{a}	11.51±0.43^{d}

Data are the mean of three replicates ± standard deviation. The same letters in the lines do not differ by the Scott-Knott test (P <0.05). Moisture content of medicinal plant flours, in g 100 g'^{1} : *Aloe* = 8.53; *Simaba* = *8*.42; *Baccharis* = *8*.56; *G. cambogia* = 3.94; *Tourneforticr.* 9.90.

Variations in circulating calcium content can affect food intake. High calcium levels decrease food intake and it is assumed that this effect is due to the greater availability of calcium for ion channels. Several studies have shown that obese patients on high-calcium diets show a reduction in body fat (HEANEY, 2003; MOORE et al., 2004; ZEMEL et al., 2004).

The bioactive compounds present in plants act to maintain health and reduce the risk of disease, but depending on the concentration, these compounds can cause damage to health, highlighting the need for studies to characterize these compounds in plant extracts. The bioactive compounds present in the MFPs used as aids in the treatment of obesity are shown in Table 4.

Phenolic compounds were found in all the plants, with *T. paniculata* having the highest content (36.19 g 100 g^{-1} DM), and *G. cambogia* (0.09 g 100 g^{-1} DM) and *A. vera* (0.15 g 100 g^{-1} DM) having the lowest. The maximum dose of phenolic compounds suggested for humans is around 1 g per day^{-1} (SCALBERT et al., 2005), so with just 3 g of *T. paniculata,* the daily limit is reached. These plants are not used for food, but for the treatment of obesity. If consumed in small quantities, only *T.*

Table 4. Content of bioactive compounds in dry matter of medicinal plants.

Medicinal plant	Phenolic compounds (gioo g')[1]	Oxalic acid (gioo g')[1]	Nitrate (g kg)[1]	Trypsin inhibitor (TIA mg')[11]	Saponins (gioo g')[1]
Aloe vera	0.15±0.02[d]	ND[2]	0.77±0.01[d]	ND	0.07±0.00[d]
Simaba femtginea	1.62±0.03[c]	0.97±0.09[a]	0.81±0.02[c]	0.17±0.02[c]	0.13±0.00[c]
Baccharis trimera	4.03±0.21[b]	0.91±0.03[a]	2.68±0.12[b]	6.26±0.32[b]	0.54±0.03[b]
Garcinia cambogia	0.09±0.01[d]	ND	0.16±0.01[e]	ND	0.07±0.00[d]
Tournefortia paniculata	36.19±0.91[a]	0.93±0.02[a]	6.96±0.13[a]	22.01±2.40[a]	1.00±0.09[a]

Data are the mean of three replicates ± standard deviation. The same letters in the columns do not differ by the Scott- test.

Knott's test (P <0.05). TIA: trypsin inhibitory activity, in nmol min'[1] mg'[1] . [2]ND: Not detected. Moisture content of medicinal plant flours, in g 100 g'[1] : *Aloe* = 8.53; *Simaba* = *8*.42; *Baccharis* = 8.56; *G. cambogia* = 3.94; *Tournefortia*. 9.90. *paniculata* may offer some kind of health risk due to its high content of phenolic compounds.

The levels of phenolic compounds in the leaves of *B. trimera* (4.03 g 100 g^{-1} DM) were higher than those observed in other studies on this plant, whose levels ranged from 0.045 to 2.67 g 100 g^{-1} DM (FREITAS et al., 2004; SOUZA et al., 2011; OLIVEIRA et al., 2012). These differences may be due to the different ways of preparing the plant (maceration and infusion) and the use of other extractants, such as ethanol, ethyl acetate, butanol, among others.

For *A. vera*, the contents exceeded those recorded by Moniruzzaman et al. (2012), which was 0.0008 g 100 g^{-1} DM. In the same study, the authors also found that *A. vera* leaves had higher phenolic contents than the gel, indicating the use of the leaves as antioxidants. On the other hand, for *G. cambogia*, the contents were lower than those reported by Subhashini, Nagarajan & kavimani (2011), which was 7.5 g pyrocatechol 100 g^{-1} DM, and also those recorded by Jantan et al. (2012) in 22 methanolic extracts from different parts (leaves, trunks, bark and fruit) of nine *Garcinia* species, with contents ranging from 0.44 to 6.28 g gallic acid 100 g^{-1} DM. The different results are probably due to the standard used in the dosage, the different extractants, the different species, the plant parts used and even the origin of the samples. For *S. ferruginea* and *T. paniculata*, no records were found in the literature on the phenolic content of these plants.

Some phenolic compounds, such as tannins, can inhibit certain digestive enzymes, such as amylase and trypsin, resulting in weight loss and helping to treat obesity (MONTEIRO et al., 2005). They also have multiple biological effects, such as antioxidant, anti-allergic, anti-inflammatory, antibacterial, antithrombotic, vasodilatory and cardioprotective (BALASUDRAM et al., 2006), showing a wide field of application for the phenolics in these plants.

There was no statistical difference between the plants *B. Trimera*, *S. ferruginea* and *T. paniculata in terms* of oxalic acid content, and it was not detected in *A. vera* and *G. cambogia*. Oxalic acid levels above 10 g are considered toxic to health; therefore, the levels of this substance found in these plants do not pose a health risk. The toxic effect of oxalic acid on the body has been associated with a reduction in the bioavailability of some essential minerals, such as calcium, and the main consequences are hypocalcemia and rickets, as well as affecting the absorption of iron, magnesium and zinc (SIENNER, 2005).

Nitrate was also found in all the plants, with levels ranging from 0.16 to 6.96 g kg^{-1} DM. The acceptable daily intake of nitrate is 5 mg kg^{-1} body weight (WHO, 2003). Excessive consumption of this compound can lead to cyanosis due to the formation of metmyoglobin and neoplasms due to the formation of N-nitroso compounds (FAQUIN; ANDRADE, 2004). Taking a 60 kg person as an example, they could ingest 300 mg of nitrate, and this value would only be reached with large quantities of the plants analyzed. Therefore, the levels of nitrate found in these medicinal plants are not capable of causing a health risk.

T. paniculata showed the highest trypsin inhibition potential (22.01 trypsin inhibitory activity in nmol min^{-1} (TIA) mg^{-1} DM), followed by *B. trimera* (6.26 TIA mg^{-1} DM), and *S. ferruginea* (TIA 0.17 mg^{-1} DM). The presence of trypsin inhibitors was not detected in the *A. vera* and *G. cambogia* plants. Souza et al. (2011) observed the presence of trypsin inhibitors in aqueous and methanolic extracts of *B. trimera* leaves, corroborating the results of this study, in which trypsin inhibition was observed for *B. trimera*; however, these authors expressed their results as a percentage of trypsin inhibition, thus making it impossible to compare with the activity observed in this study, which was expressed as TIA mg^{-1} DM. The presence of the trypsin inhibitor, especially in *T. paniculata,* shows its specificity in inhibiting proteolytic enzymes, which can lead to a decrease in protein digestibility, reducing the animals' weight gain.

Saponins were found in all the species studied, and *T. paniculata* had the highest content (1.00 g 100 g^{-1} DM). The saponin content recorded in this study for *B. trimera* (0.54 g 100 g^{-1} DM - ethanolic extract) is within the range reported by Souza et al. (2011) for this plant, which varied from 0.23 (aqueous extract) to 0.75 (methanolic extract), in g 100 g^{-1} DM. In the same study, these authors also found no hemolytic effect in tests carried out with extracts of this plant, results which indicate the low toxicity of the saponins present in these leaves.

Among the side effects that saponins can cause are changes in reproduction and growth, as well as a reduction in nutrient absorption due to changes in the permeability of cell membranes (FRANCIS et al., 2002), highlighting the need for studies to verify the toxicological potential of this phytochemical present in plant extracts.

However, saponins can help in the treatment of obesity, as they can inhibit digestive enzymes and act to reduce cholesterol in human plasma by forming micelles in the small intestine with bile acids, thus preventing their reabsorption (PEREIRA; CARDOSO, 2012), and this phytochemical may be responsible for possible anti-obesity properties in the plants analyzed in this study.

In the literature consulted, few studies were found related to the presence of the bioactive compounds determined in this work for the plants analyzed, reflecting the concern about the possible harmful effects they may cause, highlighting the need for studies with these plants.

CONCLUSION

The plants showed high levels of substances with the potential to cause weight loss, such as dietary fiber, some minerals, phenolic compounds, saponins and trypsin inhibitors. However, it is too early to recommend the use of these plants in terms of the possible risks and benefits for human health. Further studies into toxicity, efficacy and safety are needed, especially in relation to the saponins found in all the plants, and the high levels of phenolic compounds in *T. paniculata*. However, knowledge of these chemical constituents of plants contributes to their better use, both by the population and by the pharmaceutical industry, resulting in greater use and economic value.

ACKNOWLEDGMENTS

The authors would like to thank CAPES for the doctoral scholarship and FAPEMIG for the financial support.

ARTICLE 2

Inhibition of digestive enzymes by aqueous extracts of medicinal plants used to treat obesity

Simão, Anderson Assaid*, Correa, Angelita Duarte and Chagas, Pricila Maria Batista

Department of Chemistry, Biochemistry Laboratory, Universidade Federal de Lavras - UFLA, Caixa Postal 3037, CEP 37200.000, Lavras, MG, Brazil.

Published in: **Journal of Medicinal Plants Research**, v. 6, n. 47, p. 5826-5830, 2012.

SUMMARY

The aim of this study was to carry out digestive enzyme inhibition tests with aqueous extracts of the medicinal plants *Aloe vera* (L.) Burm. (aloe), *Simaba ferruginea* St. Hil. (calunga), *Baccharis trimera* (Less.) DC (carqueja), *Garcinia cambogia* Desr, *Tournefortia paniculata* Cham. (marmalade) and with the aqueous extract of the combination of these plants in the proportions 1:1.5:0.5:1.5:0.5, used as aids in the treatment of obesity. Inhibition analyses of the enzymes a-amylase, a-glucosidase, lipase and trypsin were carried out in the presence and absence of a simulated gastric fluid. In the absence of simulated gastric fluid, the enzymes were inhibited to varying degrees by the plant extracts, except for lipase, which was not inhibited at all. In the presence of simulated gastric fluid, only the enzymes a-amylase and a-glucosidase were inhibited by the plant extracts. The combination of plants did not inhibit any of the digestive enzymes evaluated. These results show that the aqueous extracts of the plants aloe, carqueja and marmalade have potential as adjuvants in the treatment of obesity and other dyslipidemias, as they inhibit a-amylase (marmalade) and a-glucosidase (aloe, carqueja and marmalade) after simulated gastric digestion; the same cannot be said of the plants in combination.

Key words: a-amylase, a-glucosidase, lipase, trypsin, medicinal plants, obesity.

INTRODUCTION

Obesity is a chronic disease resulting from the excessive accumulation of body fat, which causes damage to the health of adults, adolescents and children in both developed and developing countries, with significant losses not only in quality of life, but also in longevity.

The prevalence of obesity has been increasing at an alarming rate worldwide and has become a major health problem in modern society. Approximately 1.6 billion people are overweight, of which 400 million are obese, taking the disease to global epidemic status (Tucci et al., 2010).

Obesity causes psychological problems, frustration, unhappiness and predisposes the body to a series of diseases, in particular cardiovascular diseases, some types of cancer, diabetes and hypertension, causing increased economic costs for governments and society (Bray, 2004).

Strategies for losing weight and treating obesity generally involve a combination of dietary changes, increased physical activity, behavioral therapy, pharmacotherapy and, in extreme cases, surgery (Celleno et al., 2007). Another widely used option is natural products, which have seen a considerable increase in consumption in recent years, mainly due to the fact that people believe they are ingesting medicines that don't cause harm to their health and are inexpensive.

Research into therapeutic alternatives, mainly using medicinal plants, has been gaining ground and importance in the pharmaceutical industry, revealing itself as a very promising option for the discovery of new phytomedicines and phytotherapics, due to the large number of plant species that have not yet been studied, representing a vast field of substances to be discovered (Viegas Jr et al., 2006).

For the treatment of obesity, molecular targets such as enzymes and receptors present in natural products have been studied in order to find drugs based on the mechanism of enzyme inhibition that cause beneficial changes in metabolism, being an excellent alternative for the safe and effective development of anti-obesity drugs (Bhutani et al., 2007).

In this context, the enzymes a-amylase and a-glucosidase are responsible for processing carbohydrates, acting in the breakdown of starch and in the absorption of monosaccharides by enterocytes (Kandra, 2003; Ota et al., 2009). Inhibitors of these enzymes, present in plants, offer a promising strategy to help treat obesity, hyperglycemia associated with type 2 diabetes and hypertension, by reducing the breakdown of starch and the absorption of glucose in the intestine (Kwon et al., 2006).

In addition to these, lipase, involved in lipid metabolism, is also an interesting target for inhibitors, as its inhibition leads to a reduction in triglyceride absorption, causing a reduction in calorie expenditure and weight loss. On the other hand, the inhibition of trypsin, involved in protein digestion, unlike the other inhibitions, is characterized as having a harmful effect, as it prevents the complete absorption of amino acids, which are of fundamental importance to the body, present in food (Friedman and Brandon, 2001).

Much research has demonstrated the efficacy, importance and potential use of these enzyme inhibitors in the treatment of obesity and associated comorbidities and reinforces the need to search for new sources of these inhibitors: amylases (Obiro et al., 2008; Udani et al., 2009), glycosidases (Kwon et al., 2006) and lipases (Sharma et al., 2005; Souza et al., 2011). As such, digestive enzyme inhibitors that help to limit the intestinal absorption of carbohydrates and fats in the initial phase may prove useful as aids in the treatment of obesity.

Several medicinal plants, such as *Aloe vera* (L.) Burm. (aloe), *Simaba ferruginea* St. Hil. (calunga), *Baccharis trimera* (Less.) DC (carqueja), *Garcinia cambogia* Desr., and *Tournefortia paniculata* Cham. (marmalade) are used as aids in the treatment of obesity. However, studies on the presence of enzyme inhibitors in the extracts of these plants, which may be involved in or even responsible for their anti-obesity properties, are scarce in the literature.

Based on the above, the aim of this work was to carry out digestive enzyme inhibition tests with aqueous extracts of aloe, calunga, carqueja, *G. cambogia* and marmalade and with the aqueous extract of the combination of these plants, so that these plants can be used as aids in the treatment of obesity.

MATERIALS AND METHODS

Sample collection and preparation

The plants *Simaba ferruginea* St. (calunga), *Baccharis trimera* (Less.) DC (carqueja) and

Tournefortia paniculata Cham. (marmelinho) were purchased at the municipal market in Belo Horizonte, Minas Gerais, Brazil, and transported to the Biochemistry Laboratory of the Chemistry Department of the Federal University of Lavras (UFLA). The marmalade and carqueja leaves were washed in running water and distilled water and then placed together with the bark obtained from the calunga trunk in a forced ventilation oven to dry for 48 hours at temperatures ranging from 30°C to 35°C. After drying, the leaves and bark were ground in a Willey-type mill and the flours stored in hermetically sealed jars until analysis. The commercial powder of *Aloe vera* (L.) Burm. (aloe) (mucilage) and *Garcinia cambogia* Desr. (fruit) were purchased from FLORIEN, a distributor of pharmaceutical products.

The plant flours were mixed to make a simulated herbal medicine, from the combination of aloe, calunga, carqueja, G. *cambogia* and marmalade in the ratio 1:1.5:0.5:1.5:0.5, respectively; the same combination is used to make the herbal medicine known by the trade name 'Moder Diet'.

Determining humidity

Moisture determination was carried out in triplicate on the medicinal plant flours (MPF) according to the method of the Association of Official Analytical Chemists - AOAC (2005), which consists of the loss of water by dehydration, at temperatures varying between 100°C and 105°C.

Preparing the extract

The MPFs were mixed with distilled water in a ratio of 1:25 (w/v) and placed in a horizontal shaker at room temperature for 1 hour. The mixture was then filtered through filter paper and used as inhibitors in the enzymatic analyses.

Obtaining enzymes

For the assays, porcine pancreatic a-amylase type VI (SIGMA), porcine pancreatic trypsin and porcine lipase type II (MERCK) were used. The a-glucosidase was obtained from fresh porcine duodenum supplied by the Department of Animal Sciences at UFLA, which was ground in a blender with 0.5 mol L Tris-HCl buffer^{-1} , pH 8.0 at 4°C, to extract the enzymes from the enterocyte membranes and processed in a mixer until completely homogenized. The homogenate was filtered through nylon mesh and centrifuged for 10 minutes at 2,500 x g at 4°C. The supernatant was collected and used as an enzyme extract (Souza et al., 2011). **a-amylase activity**

The activity of a-amylase was determined according to the methodology proposed by Noelting and Bernfeld (1948). Thus, 50 pL of the plant extracts and 50 pL of a-amylase were pre-incubated for 20 minutes in a 37°C water bath. The substrate was 1% starch prepared in 0.05 mol L Tris buffer^{-1} , pH 7.0 with 38 mmol L^{-1} NaCl and 0.1 mmol L^{-1} CaCl2. After adding 100 pL of the substrate, the mixture was incubated for four time periods. The reaction was stopped by adding 200 pL of 3,5 dinitrosalicylic acid and the product was read in a spectrophotometer at 540 nm.

a-glucosidase activity

The activity of a-glucosidase was determined according to Kwon et al. (2006), using 5 mmol L^{-1} p-nitrophenyl-a - D- glucopyranoside in a 0.1 mol L citrate-phosphate buffer^{-1} pH 7.0 as the substrate. In the assay, 50 pL of the plant extracts and 100 pL of enzyme were incubated in a water bath at 37°C for four periods of time, after the addition of 50 pL of substrate. The reaction was stopped by adding 1,000 pL of 0.05 mol L NaOH^{-1} and the product was read in a spectrophotometer at 410 nm.

Lipase activity

In each analysis, the mixture of 100 pL of lipase, 50 pL of the plant extracts and 50 pL of the substrate *p-nitrophenyl* laurate 4 mmol L^{-1} in Tris-HCl buffer 0.05 mmol L^{-1} pH 8.0 containing 0.5% Triton-X100 was incubated for four periods of time. The reaction was stopped by transferring the tubes to an ice bath and adding 1,000 pL of 0.05 mmol L Tris-HCl buffer^{-1} pH 8.0. The yellow-colored *p-nitrophenol*, a product of the action of lipase on *p-nitrophenyl* palmitate, was read in a spectrophotometer at 410 nm (Souza et al., 2011).

Trypsin activity

Trypsin activity was determined according to the methodology proposed by Erlanger (1961); in which 200 pL of the plant extracts and 200 pL of enzyme were incubated in a water bath at 37°C for four periods of time, after adding 800 pL of the substrate *p-benzoyl-DL-arginine-p-nitroanilide*

(BApNA) prepared in TRIS (tris(hydroxymethyl)aminomethane) buffer 0. 05 mol L , TRIS (tris(hydroxymethyl)aminomethane) buffer pH 8.2 with 20 mmol L CaCl2.05 mol L^{-1} , TRIS (tris(hydroxymethyl)aminomethane) buffer pH 8.2 with 20 mmol L^{-1} CaCl2. The reaction was stopped by adding 200 pL of 30% acetic acid and the product was read in a spectrophotometer at 410 nm.

Determining inhibition

Enzyme inhibition was obtained by determining the slopes of the lines (absorbance x time) of the activity tests of the control enzymes (without plant extract) and the enzymes + inhibitor (with plant extracts). The slope of the line is due to the rate of product formation per minute of reaction and the presence of the inhibitor causes this slope to decrease. From this slope, the absorbance values were converted into pmol of product using a glucose standard curve for amylase and *p-nitrophenol* for glycosidase and lipase, while for trypsin, the molar extinction coefficient of BApNA determined by Erlanger (1961) was used.

Preparation of simulated gastric fluid

In order to simulate the digestion process *in the* stomach *in vitro*, enzyme activity tests were also carried out in the presence of a simulated gastric fluid. To this end, the plant extracts were incubated with the simulated gastric fluid, prepared according to The United States Pharmacopeia - USP (1995), for 1 hour in a 37°C water bath. After this period, they were neutralized with sodium bicarbonate salt to physiological pH and only then were the inhibitory activity tests carried out.

RESULTS AND DISCUSSION

The inhibition of digestive enzymes is a promising alternative for the treatment of obesity, mainly due to the fact that they act on the small intestine, without acting on the central nervous system, where the usual anorexiants act. The results of enzyme inhibition by the aqueous extracts of medicinal plants are shown in Table 1.

The a-amylase was only inhibited by the marmalade extract, which showed an inhibition potential of 2,907.13 pmol min^{-1} g^{-1} dry

Table 1. Inhibition of digestive enzymes by aqueous extracts of medicinal plants, expressed in enzyme inhibition units (UIE[1]).

Medicinal plant	a-amylase	a-glucosidase	Lipase	Trypsin
Aloe	nd[2]	1.23±0.05	nd	nd
Calunga	nd	nd	nd	nd
Carqueja	nd	0.58±0.03	nd	10.38±0.81
Garcinia cambogia	nd	nd	nd	nd
Marmelinho	2,907.13±7.64	35.46±0.58	nd	176.68±7.05
Phytotherapeutic[0]	nd	nd	nd	nd

Data are the mean of triplicates ± standard deviation. Moisture content of medicinal plant flour in g 100 a'[1] : aloe = 8.53; calunga = 8.42; carqueja = 8.56; *Garcinia cambogia* = 3.94; marmalade = 9.90. [f]One UIE is equal to 1 pmol min'[1] g'[1] of dry matter.[2] nd: inhibition not detected. [3]Herbal medicine: made from a combination of aloe, calunga, carqueja, *Garcinia cambogia* and marmalade in the ratio 1:1.5:0.5:1.5:0.5, respectively.

matter - MS. This potential is considered high when compared to Pereira et al. (2010), who analyzed different extraction conditions (solvents, times and temperatures) for a-amylase inhibitors in white beans and found inhibition between 36.88 and 66.89 pmol min^{-1} g^{-1} DM. Souza et al. (2011), analyzing the inhibitory potential of different carqueja extracts, also found no inhibition of a-amylase.

A-glucosidase was inhibited by the extracts of aloe, carqueja and marmalade, with the latter plant showing an inhibition potential of 35.46 pmol min^{-1} g^{-1} DM, which is much higher than the other plants that inhibited a-glucosidase. The inhibitory potential of marmalade in this study exceeds that found by Pereira et al. (2011), who, analyzing commercial samples of *Hoodia gordonni,* a plant used as an aid in the treatment of obesity, found inhibition between 10.40 and 16.70 pmol min^{-1} g^{-1} DM. However, the potentials of aloe and carqueja are lower than those of Hoodia *gordonni*. Inhibition of these enzymes induces tolerance to carbohydrates, prolongs gastric emptying, causes satiety and weight loss; all these effects may be useful in the treatment of obesity (Chen et al., 2008).

Inhibition of the lipase involved in lipid metabolism was not detected for any of the aqueous extracts of the medicinal plants analyzed. Studies in the literature show that lipase inhibitors have been isolated in the alcoholic extracts of plants, mainly in methanol (Sharma et al., 2005; Sugimoto et al., 2009). These studies suggest that organic compounds soluble in methanol have some structural characteristics that have the ability to bind and inhibit pancreatic lipase.

A high level of inhibition was observed for trypsin: 176.68 pmol min^{-1} g^{-1} DM, (marmalade) and 10.38 (carqueja). Trypsin inhibitors present in the diet can cause a reduction in the growth rate of animals, accompanied by a decrease in protein digestibility, which can lead to weight loss (MacDougall et al., 2005). The combination of plants did not inhibit any of the digestive enzymes analyzed.

When the medicinal plant extract is ingested, passage through the gastrointestinal tract can lead to structural changes in the inhibitors due to the acidic pH of the stomach, causing them to be deactivated. Considering the significant inhibition of a-amylase, a-glucosidase and trypsin in the presence of the plant extracts, they were subjected to a test of the possible action of gastric fluid on the inhibitory activity of the extract (Table 2).

It was observed that there was a 14% reduction in the inhibitory potential of the marmalade extract on a-amylase in the presence of simulated gastric fluid. For a-glucosidase, the simulated fluid caused a reduction in the inhibitory potential of the plant extracts of 60% (aloe), 32.8% (carqueja) and 27% (marmalade). However, even with the reduction in the enzyme inhibition potential of the plant extracts, the results were still significant. The carqueja and marmalade extracts lost their activity on the trypsin enzyme, which indicates that the trypsin inhibitors present in these extracts are not resistant to passage through the gastric fluid, which can lead to modifications in the inhibitor molecule due to the acidic pH of the stomach or the presence of proteinases, inactivating it.

The inhibition results suggest that aloe, carqueja and especially marmalade may be a good source of a-amylase and a-glucosidase inhibitors, which can be used as aids in

Table 2. Inhibition of digestive enzymes by aqueous extracts of medicinal plants after exposure to simulated gastric fluid, in enzyme inhibition units (UIE[1]).

Medicinal plant	a-amylase	a-glucosidase	Lipase	Trypsin
Aloe	nd[2]	0.49±0.02	nd	nd
Calunga	nd	nd	nd	nd

Carqueja	nd	0.39±0.04	nd	nd
Garcinia cambogia	nd	nd	nd	nd
Marmelinho	2,512.55±8.54	25.90±1.12	nd	nd
Phytotherapeutic[3]	nd	nd	nd	nd

Data are the mean of triplicates ± standard deviation. Moisture content of medicinal plant flour in g 100 a'[1] : aloe = 8.53; calunga = 8.42; carqueja = 8.56; *Garcinia cambogia* = 3.94; marmalade = 9.90. [r]One UIE is equal to 1 pmol min'[1] g'[1] of dry matter.[2] nd: inhibition not detected. [3]Herbal medicine: made from a combination of the plants aloe, calunga, carqueja, *Garcinia cambogia* and marmalade in the ratio 1:1.5:0.5:1.5:0.5, respectively. treatment of obesity which can be caused by a high level of carbohydrates in the diet. The resistance of the inhibitors to passing through simulated gastric fluid is a strong indication that these results will be repeated in *in vivo* trials.

CONCLUSIONS

The aqueous extracts of aloe, carqueja and marmalade plants have potential as adjuvants in the treatment of obesity and other dyslipidemias, as they inhibit a-amylase (marmalade) and a-glucosidase (aloe, carqueja and marmalade) after simulating gastric digestion; the same cannot be said of plants in combination (herbal medicines).

ACKNOWLEDGMENTS

We would like to thank CAPES for the PhD scholarship and FAPEMIG for the financial support and scientific initiation scholarship.

ARTICLE 3

Antioxidants in medicinal plants used to treat obesity

Anderson Assaid Simão[1]* , Fabiola F. Lage[1] , Pricila M. B. Chagas[1] , Rodrigo M. Fraguas[1] , Juliana M. Freire, Tamara R. Marques[1] and Angelita D. Correa *[1]

[1]Department of Chemistry, Biochemistry Laboratory, Federal University of Lavras - UFLA, Caixa Postal 3037, CEP 37200.000, Lavras, MG, Brazil.

Published in: **European Journal of Medicinal Plants,** v. 3, n. 3, p. 429-443, 2013.

SUMMARY

Objectives: The aim of this work was to quantify the phenolic compounds, flavonoids, vitamin C, total carotenoids, p-carotene and lycopene and to measure the antioxidant activity of the medicinal plants *Aloe vera* (L.) Burm. (aloe), *Simaba ferruginea* St. Hil. (calunga), *Baccharis trimera* (Less.) DC (carqueja), *Garcinia cambogia* Desr., and *Tournefortia paniculata* Cham. (marmalade) and of

* *Corresponding author: E-mail: andersonbsbufla@yahoo.com.br; angelita@dqi.ufla.br;*

the herbal medicine made with the combination of these plants.
Place and duration of the study: Chemistry Department of the Federal University of Lavras - UFLA, Brazil, between June 2011 and September 2012.
Methodology: Phenolic compounds, flavonoids, vitamin C, total carotenoids and p-carotene and lycopene levels were quantified by UV-Vis spectrophotometry and antioxidant activity by the ABTS and p-carotene/linoleic acid methods.
Results: High levels of phenolic compounds were found in marmalade (36.19 g 100 g^{-1} dry matter - DM), followed by carqueja (4.03 g 100 g^{-1} DM) and calunga (1.62 g 100 g^{-1} DM); flavonoids in marmalade (480.30 mg 100 g^{-} 1DM) and carqueja (173.68 mg 100 g^{-1} DM); vitamin C in marmalade (652.80 mg 100 g^{-1} DM) and *G. cambogia* (127.63 mg 100 g^{-1} DM); and carotenoids in marmalade (23.16 mg 100 g^{-1}). The antioxidant activity, in pmol trolox g^{-1} , using the ABTS method, was considered to be moderate in the aqueous (728.80) and ethanolic (731.06) extracts of marmalade, and weak for the other plants. However, by the p-carotene/linoleic acid method, the aqueous and ethanolic extracts of marmalade showed great antioxidant potential at all the concentrations tested (above 80% inhibition), and those of carqueja, calunga and the ethanolic of the herbal medicine, at concentrations of 40,000 and 20,000 mg L^{-1} , also showed good antioxidant potential (above 60% inhibition).
Conclusion: These five plant species showed antioxidant activity with potential for use in pharmaceutical and food preparations, with possible health benefits.
Keywords: Phenolic compounds; Flavonoids; Vitamin C; Carotenoids; Antioxidant potential.

1. INTRODUCTION

In recent years, the effects of antioxidants have been investigated in relation to diseases. Research has attempted to explain the benefits of these substances for the prevention and treatment of various types of disease [1,2].

Antioxidants are substances that fight free radicals, which are extremely reactive species that cause the oxidation of various biomolecules (lipids, proteins and nucleic acids) present in our bodies, causing various pathologies such as cancer, neurodegenerative disorders, cardiovascular diseases, diabetes and other chronic diseases, and free radicals can be the cause or aggravating factor of the general condition of these diseases [3]. Research is therefore looking for alternatives to reduce the harmful effects of free radicals and improve the body's antioxidant capacity, as a way of treating and preventing diseases and their complications.

In view of the epidemiological growth of these diseases, different foods and plants are being studied for their substances capable of neutralizing the effects of free radicals, such as phenolic compounds and vitamin C (hydrophilic antioxidants), vitamin E and carotenoids (lipophilic antioxidants), and certain minerals such as zinc and selenium [4].

In this context, plants that are popularly known to have some therapeutic purpose, and which have not been the subject of studies to prove their effects, have become the central objective of research aimed at developing new drugs to help treat diseases.

Aloe vera (L.) Burm. (aloe), *Simaba ferruginea* St. Hil. (calunga), *Baccharis trimera* (Less.) DC (carqueja), *Garcinia cambogia* Desr., *Tournefortia paniculata* Cham. (marmalade) are plants used, alone or in combination, as aids in the treatment of obesity [5,6]. When they are associated, they are used in an attempt to obtain better results, as is the case with the herbal medicine Moder diet, made from a combination of these plants to help treat obesity, with few studies related to its antioxidant properties which, if proven, could help treat various other diseases, some directly related to obesity, such as diabetes, cardiovascular diseases, hypertension, among others, which can be caused or aggravated by free radicals [7].

Based on the above, the aim of this work was to quantify the phenolic compounds, flavonoids, vitamin C, total carotenoids, p-carotene and lycopene and measure the antioxidant activity of the medicinal plants aloe, calunga, carqueja, *G. cambogia* and marmalade and the herbal medicine made from a combination of these plants, with the aim of evaluating the possible use of these plants in the fight against free radicals and consequently in the treatment of various diseases caused by them.

2. MATERIAL AND METHODS

2.1 Sample collection and preparation

Baccharis trimera (Less.) DC (carqueja) and *Tournefortia paniculata* Cham. (marmelinho), and the bark from the trunk of *Simaba ferruginea* St. (calunga) were purchased at the municipal market in Belo Horizonte, Minas Gerais, Brazil, in January 2011, in three repetitions, and transported to the Biochemistry Laboratory of the Chemistry Department at the Federal University of Lavras (UFLA). The marmalade and carqueja leaves were washed in running water and distilled water and then placed together with the bark obtained from the calunga trunk in a forced ventilation oven to dry for 48 hours at temperatures ranging from 30°C to 35°C. After drying, the leaves and bark were ground in a Willey mill and the flours stored in hermetically sealed jars until analysis. The commercial powder of *Aloe vera* (L.) Burm. (aloe) (mucilage) and *Garcinia cambogia* Desr. (fruit) were purchased from FLORIEN, a distributor of pharmaceutical products. The aloe powder was obtained by freeze-drying the plant's mucilage, while the *G. cambogia* powder *was* spray-dried.

The plant flours were mixed to make a simulated herbal medicine, using a combination of aloe, calunga, carqueja, *G. cambogia* and marmalade in proportions of 1:1.5:0.5:1.5:0.5, respectively; the same combination is used to make the herbal medicine known by the trade name 'Moder Diet'.

2.2 Determining humidity

Moisture was determined in the medicinal plant flours according to the method of the Association of Official Analytical Chemists - AOAC [8], which consists of the loss of water by dehydration at temperatures ranging from 100°C to 105°C.

2.3 Phenolic compounds

The phenolic compounds were extracted using 1 g of sample in 50 mL of 50% methanol under reflux for three consecutive times at 80°C. The extracts were collected, evaporated to 25 mL and the phenolic compounds were measured using the Folin-Denis reagent and tannic acid as a standard [8].

2.4 Total flavonoids

Total flavonoid contents were measured using the same extracts as the phenolic compound analyses, using the aluminum chloride colorimetric method, with catechin used as a standard [9].

2.5 Vitamin C

Ascorbic acid was extracted using 0.5 g of the sample in 50 mL of oxalic acid and 0.1 g of diatomaceous earth, stirring for 15 minutes. After filtration (Whatman N° 40), the vitamin C in the extract was measured using ascorbic acid as a standard [10].

2.6 Total carotenoids

To determine total carotenoids, extraction was carried out according to Higby [11], using 0.5 g of sample in 40 mL of a 3:1 isopropyl alcohol:hexane extractant solution. The contents were transferred to a 125 mL separating funnel wrapped in aluminum, where the volume was topped up with distilled water. It was left to stand for 30 minutes, followed by washing the material and discarding the aqueous phase. This operation was repeated three more times. The contents were filtered through absorbent cotton sprayed with 99% anhydrous sodium sulphate into a 25 mL volumetric flask wrapped in aluminum foil, where 5 mL of acetone was added and the volume topped up with hexane. Readings were taken at 450 nm and the results expressed in mg 100 g^{-1}, calculated using the formula:

Total carotenoids = (A450 x 100)/(250 x L x W), where:

A_{450} = absorbance, L = cuvette width in cm and W = ratio between the mass of the original sample and the final volume of the dilution in ml.

2.7 p-carotene and lycopene

For the determination of p-carotene and lycopene, the same extract was used as for the analysis of total carotenoids, in which these extracts were taken for absorbance readings in a spectrophotometer at four wavelengths: (453, 505, 645 and 663 nm) [12]. The following equations were used to calculate the concentrations of p-carotene and lycopene:

β-carotene (mg 100 g^{-1}) = $0.216\ A_{663} - 1.22\ A_{645} - 0.304\ A_{505} + 0.452\ A_{453}$. lycopene (mg 100 g^{-1}) = $- 0.045\ A_{663} + 0.204\ A_{645} + 0.372\ A_{505} - 0.0806\ A_{453}$.

2.8 Antioxidant activity

2.8.1 Preparing the extract

The antioxidants were extracted using two extractants: water (1:25, w/v) and ethanol (1:25, w/v). For each of the extractants, the samples were kept stirring for 1 hour and then filtered through filter paper. All the extractions were carried out in three repetitions, protected from light and then submitted to the antioxidant activity (AA) detection process using the methods described below.

2.8.2 p-carotene/linoleic acid method

Dilutions of 20,000 and 10,000 mg L^{-1} were prepared from the crude extracts of the samples (40,000 mg L^{-1}). The methodology used was developed by Rufino et al. [13], with modifications.

To prepare the p-carotene/linoleic acid solution system, 50 pL of p-carotene diluted in chloroform (20 g L^{-1}) were used, to which 40 pL of linoleic acid, 530 pL of tween 20 (emulsifier) and 1 mL of chloroform were added for solubilization. In a flask covered with aluminum foil to protect against light, the chloroform was evaporated in a rota-evaporator and 100 mL of oxygen-saturated water (distilled water treated with oxygen for 30 minutes) was added, and the combination was stirred until the solution system was yellow-orange in color. In test tubes, 2.5 mL of this solution system was added to 0.2 mL of each dilution of the sample used for the test. Control tubes were made containing 2.5 mL of the solution system with 0.2 mL of BHT (butylated hydroxytoluene - synthetic antioxidant), quercetin and rutin, which are flavonoids with proven antioxidant action, all at a concentration of 200 mg L^{-1} . In laboratory tests, the concentration of 200 mg L^{-1} of BHT was found to provide the greatest protection to the system when compared to the others; therefore, its use is suggested. After homogenization, the readings were taken on a spectrophotometer at 470 nm, using water to calibrate the spectrophotometer, which was taken as the reading at time zero (initial). The tubes were placed in a water bath at 40°C and the readings were taken after 2 hours.

2.8.3 ABTS method

The methodology used was developed by Re et al. [14], with modifications by Rufino et al. [15]. Four different dilutions were made of the extracts obtained for the tests and subsequent construction of the analytical curves.

Analytical curves were made with trolox (6-hydroxy-2,5,7,8-tetramethylchroman-2-carboxylic acid) and ascorbic acid, as well as comparison tests with the standards BHT, rutin and quercetin, prepared at a concentration of 200 mg L^{-1} .

2.9 Statistical analysis

The data are the mean of three replicates ± standard deviation analyzed by one-way analysis of variance (ANOVA) and when this analysis showed a significant difference, the Skott-Knott test ($P<0.05$) was used for the comparison of means. All statistical tests were carried out using the R statistical software (version 2.15.2) [16].

3. RESULTS AND DISCUSSION

3.1 Antioxidant substances

The antioxidant contents of the medicinal plant flours and the herbal medicine are shown in Table 1. Phenolic compounds were found in all the plants, with marmalade having the highest content (36.19 g 100 g^{-1} dry matter - DM) and *G. cambogia* the lowest (0.09 g 100 DM g^{-1} *)*. The content of phenolic compounds in the herbal medicine was considered to be relatively high (4.56 g 100 DM g^{-1}) and was only lower than that of the marmalade.

The levels of phenolic compounds in carqueja leaves (4.03 g 100 g^{-1} DM) were higher than those observed in other studies on this plant, whose levels ranged from 0.045 to 2.67 g 100 g^{-1} DM (5,17,18). These differences may be due to the different ways of preparing the plant (maceration and infusion) and the use of other extraction agents, such as ethanol, ethyl acetate, butanol, among others. For aloe, the levels exceeded those recorded by Moniruzzaman et al [19], which was 0.0008 g 100 g^{-1} DM. In the same study, the authors also found that

Table 1. Levels of antioxidants in dry matter from medicinal plants and herbal medicine.

Plants	Phenolic compounds (g100 g')[1]	Flavonoids (mg 100 g')[1]	Vitamin C (mg 100 g')[1]	Total Carotenoids (mg 100 g')[1]	ß-carotene (mg 100 g')[1]	Lycopene (mg100 g')[1]
Aloe vera	0.15±0.01 d	ND	13.37±0.60 f	0.54±0.11 e	ND	0.52±0.01 d
Calunga	1.62±0.03 c	55.75±2.76 d	44.40±0.96 d	1.50±0.02 d	ND	0.34±0.04 d
Carqueja	4.03±0.21 b	173.68±3.60 b	19.27±0.38 e	13.67±0.74 b	3.59±0.50 b	1.97±0.15 b
Garcinia cambogia	0.09±0.00 d	ND	127.63±0.70 b	0.21±0.01 e	ND	0.20±0.08 d
Marmelinho	36.19±0.91 a	480.30±4.73 a	652.80±8.66 a	23.16±0.52 a	10.09±0.94 a	3.28±0.36 a
Phytotherapy[2]	4.56±0.20 b	82.10±3.70 c	121.50±1.10 c	4.30±0.70 c	1.36±0.01 c	0.79±0.01 c

Data are the mean of three replicates ± standard deviation. The same letters in the columns do not differ by the Scott-Knott test ($P<0.05$). Moisture content of medicinal plant flour in g 100 g'^{1} : Moe vera = 8.53; calunga = 8.42; carqueja = 8.56; Garcinia cambogia = 3.94; marmalade = 9.90. [1] *Nd: Not detected.* [2]*Herbal medicine: made from a combination of Aloe vera, calunga, carqueja, Garcinia cambogia and marmalade in the ratio 1:1.5:0.5:1.5:0.5, respectively, these data being obtained by estimation.*

aloe leaves have higher phenolic contents than the gel, which indicates the use of aloe leaves as antioxidants. For *G. cambogia*, the contents were lower than those recorded by Subhashiniet al. [20], which was 7.5 g pyrocatechol 100 g^{-1} DM, and also those recorded by Jantan et al. [21] in 22 methanolic extracts from different parts (leaves, branches, bark and fruit) of 9 species of *Garcinia*, with contents ranging from 0.44 to 6.28 g of gallic acid 100 g^{-1} DM, and in both studies the Folin-Ciocateu method was used. These different results are probably due to the standard used in the dosage, the different extractants, the different species, the plant parts used and even the origin of the samples. For calunga and marmalade, no records were found in the literature on the phenolic content of these plants, so the comparison will be made with other medicinal plants. The levels of phenolic compounds in marmalade (36.19 g 100 DM g^{-1}) were very high, and higher than those reported by Ghimire et al. [22] who, in 24 medicinal plants from Nepal, found phenolic compounds, in g 100 DM g^{-1} , between 2.38 (*Drymaria cordata*) and 32.12 (*Amamum subulatum*); by Wojdylo et al. [23], who found levels between 0.07 (*Carum carvi)* and 15.15 (*Echinacia purpurea)* in 32 Polish herbs; and by Gan et al. [24], who found levels between 0.04 (*Curcuma* aromatica) and 7.57 (*Sanquisorba officinalis) in* 40 medicinal plants. The phenolic compounds in marmalade were exceeded only by *Acacia catechu* Willd with 41.47 g 100 DM g^{-1} , in a study of 133 medicinal plants from 64 different families in India [25], showing the high levels of phenolics in marmalade, which could represent several possible fields for the application of these phenolics, adding value to this plant.
Phenolic compounds act as antioxidants due to their redox properties, which allow them to act as reducing agents, hydrogen donors and metal chelators. In addition to their role as antioxidants, these compounds have a wide range of medicinal properties, such as anti-allergic, anti-inflammatory, antibacterial and antithrombotic, as well as cardioprotective and vasodilator effects [26], showing a broad field of application for phenolics in these plants.
Marmalade had the highest flavonoid content (480.30 mg 100 DM g^{-1}), followed by carqueja (173.68 mg 100 DM g^{-1}) and calunga (55.75 mg 100 DM g^{-1}). Flavonoids were not detected in either aloe or *G. cambogia.* However, for the herbal medicine, we estimated a flavonoid content of 82.10 mg 100 DM g^{-1} .
The flavonoids in carqueja were higher than those found by Borella and Fontoura [27], who recorded levels of between 12.43 and 47.50 mg of rutin 100 g^{-1} MS in 8 commercial samples of carqueja leaves. This difference is probably due to the standard used, as well as factors related to cultivation, fertilization, place of collection, age of the plant, among others. Studies with aloe [19] and *G. cambogia* [20] have shown the presence of flavonoids in the extracts of these plants, unlike the present study, in which no flavonoids were detected. These differences may be due to various factors, such as the environment, the place where the extracts were collected, prepared and handled; in addition, adulteration and/or falsification may have occurred during this process.
The flavonoid content of marmalade was higher than that recorded by Sumazian et al. [28] in the leaves of 6 Malaysian vegetables whose content, in mg 100 DM g^{-1} , ranged from 42 (*Whitania somnifera*) to 405 (*Curcuma tames*), to that of various parts of medicinal plants, in 100 g DM: *Alcea kurdica* flowers (22 mg), *Valeriana officinalis* root (110 mg), *Stachys lavandulifolium* flowers (402 mg), and lower in relation to the leaves of *Lavandula officinalis* (618 mg) and *Melissa officinalis (*1.000 mg) [29]. For the leaves of 11 plants analyzed by Djeridane et al. [4], the flavonoids of the marmalade surpassed those of 6 plants with contents between 162 (*Ruta montana*) and 454 mg 100 DM g^{-1} (*Globularia alypum*) and are lower than those reported for 24 medicinal plants from Nepal which showed contents, in mg 100 DM g^{-1} , between 1,353 (*Withania somnifera) and* 10,033 (*Artemesia vulgaris)* [22].
The consumption of flavonoid-rich foods and plants is associated with a reduction in the risk of various chronic diseases, and their protective effect is partly due to their antioxidant properties and their ability to reduce oxidative stress [30]. Epidemiological data confirms a significant relationship between a high intake of flavonoids and a reduction in carcinogenic risk, cardiovascular disease, myocardial infarction and total LDL concentrations [31,32].
Vitamin C was also found in all the plants analyzed, with levels ranging from 13.37 to 652.80 mg 100 DM g^{-1} . The estimated vitamin C content of the herbal medicine (121.50 mg 100 DM g^{-1}) was

only lower than that of *G. cambogia* and marmalade. With the discovery of the antioxidant action of this vitamin, the intake of substances with a high vitamin C content has been recommended. Therefore, these plants appear to be good sources of vitamin C with potential for use as antioxidants. The total carotenoid content ranged from 0.20 to 23.16 mg 100 DM g^{-1} . The highest amount of total carotenoids, in mg 100 DM g^{-1} , was recorded in marmalade (23.16), followed by carqueja (13.67) and the herbal medicine (4.30). *G. cambogia* had the lowest carotenoid content. Carotenoids act as lipophilic antioxidants and together with vitamin C and phenolic compounds (hydrophilic antioxidants) form a strong defense against free radicals, as they act in different cellular compartments.

p-carotene was not detected in the aloe, calunga and *G. cambogia* plants, while lycopene was found in all the plants, with marmalade and carqueja showing the highest levels of these substances. The estimated content of p-carotene (0.79) and lycopene (1.36) in the herbal medicine was only lower than that of marmalade and carqueja.

Studies by Zhao et al [33] indicate that carotenoid supplementation reduces DNA damage and that the combination of carotenoids (p-carotene and lycopene) ingested and achieved via the diet or in high doses of an individual carotenoid (12 mg) could protect against DNA damage. P-carotene is highly fat-soluble and widely transported with LDL cholesterol (75%) and HDL cholesterol (25%). It has an antioxidant function as a free radical scavenger, to reduce the extent of cell nucleus damage and to inhibit lipid peroxidation induced by free radical enzyme sources such as xanthine oxidase [34].

3.2 Antioxidant activity

The results of (AA) by the ABTS method, calculated by the trolox analytical curve and by the vitamin C standard curve, in the flours of the medicinal plants, the herbal medicine and the three standards are shown in Figure 1. Of the plants analyzed, marmalade had the highest antioxidant potential, probably due to the higher levels of phenolics, flavonoids, vitamin C and carotenoids in the leaves. For this plant, there was practically no difference in AA between the aqueous and ethanolic extracts, in trolox equivalents, or in vitamin C. For calunga, carqueja and the herbal medicine, the aqueous extracts showed higher AA than the ethanolic extracts, both for trolox equivalents and vitamin C, which can be explained by the fact that the ABTS test shows better results when in the presence of hydrophilic antioxidants. Aloe and *G. cambogia* did not show AA by the ABTS method, probably due to the low content of phenolic compounds and the absence of flavonoids in the extracts of these plants, which are substances with a strong action in capturing the ABTS radical.

The low antioxidant potential shown by the herbal medicine in the ABTS test may be due to the lower proportion of marmalade in the herbal medicine (10%), since this plant showed good antioxidant potential, and also to a negative synergistic action in the combination of these plants.

It was observed that, when compared to the standards BHT and rutin, the antioxidant potential, in trolox equivalents and vitamin C, of the aqueous and ethanolic extracts of marmalade reached 58% and 54%, on average, of the potential of those standards. In relation to quercetin, the potential of marmalade was much lower, with only 9.7% of its potential (in trolox equivalents, as well as vitamin C). The antioxidant potential of the extracts from the other plants and the herbal medicine was considered low. However, as the antioxidant substances in the quinces extracts are not found in isolated form as in the standards, this average potential of approximately 56% can be considered a good antioxidant potential.

The good antioxidant potential demonstrated by marmalade leaves is evident when compared with other studies in which, regardless of the extract,

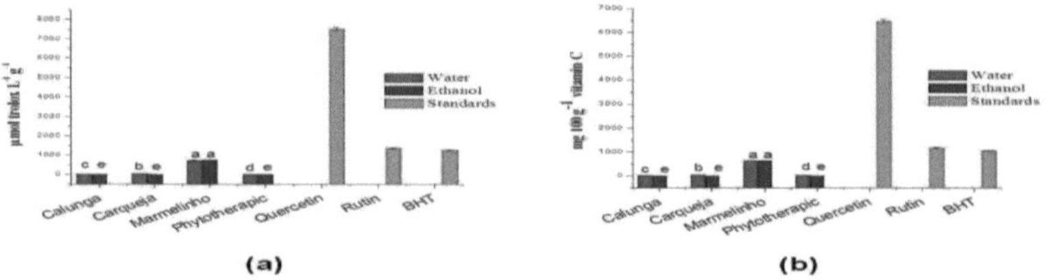

Fig. 1. Antioxidant activity by the ABTS method of aqueous and ethanolic extracts of medicinal plants, herbal medicine and three standards, expressed as (a) pmol trolox L'^{1} g'^{1} and (b) mg g'^{1} of vitamin C.

Data are the mean of three replicates ± standard deviation. Equal letters in the columns do not differ according to the Scott-Knott test (P<0.05). Herbal medicine: made from a combination of Aloe vera, Calunga, Carqueja, Garcinia cambogia and Marmalade
in a ratio of 1:1.5:0.5:1.5:1.5:0.5 respectively. exceeded that of Wojdylo et al. [23] who, in 32 Polish herbs, found potentials of between 0.0045 (Archangelica officinalis) and 3.46 (Syzygium aromaticum) pmol trolox g^{-1} *. It also exceeded that detected by Bouayed et al. [29] in various parts of medicinal plants, in mg* g^{-1} *vitamin C: 2.8 (Alcea kurdica flowers), 7.36 (Valeriana officinalis root), 15.4 (Stachys lavandulifolium flowers), 19.2 (Lavandula officinalis leaves) and 19.3 (Melissa officinalis leaves). It is exceeded only by 11 of the 132 Indian medicinal plants analyzed by Surveswaran et al. [25] and by two of the 40 medicinal plants studied by Gan et al. [24]. These results were probably due to the presence of different antioxidants, forms and extractants used in the preparation of these plants, which implies a greater antioxidant potential.*

The results of the inhibition of lipid oxidation by the p-carotene/linoleic acid method for the MPF, the herbal medicine and the three standards after 2 hours of reaction are shown in Table 2. All the plants showed potential for inhibiting lipid oxidation, except for the ethanolic extract of *G. cambogia*. Marmalade was the plant with the highest antioxidant potential in the concentrations analyzed, and at concentrations of 40,000 and 20,000 mg L^{-1} , the aqueous and ethanolic extracts had practically the same antioxidant potential, while at a concentration of 10,000 mg L^{-1} , the ethanolic extract had the highest potential.

The extracts (aqueous and ethanolic) of calunga and carqueja, at concentrations of 40,000 and 20,000 mg L^{-1} , also showed good antioxidant potential (inhibition of over 60%). The herbal medicine showed great antioxidant potential, especially its ethanolic extract, with an inhibition potential at concentrations of 40,000 and 20,000 mg L^{-1} , which was only inferior to the antioxidant potential of marmalade, showing that the combination of plants has a good antioxidant potential.

The p-carotene/linoleic acid method shows a better response for antioxidants with an apolar character. As marmalade had the highest levels of carotenoids (lipophilic antioxidants) and other antioxidants, it had the highest antioxidant potential among the plants analyzed. For calunga, carqueja and the herbal medicine, the greater potential for inhibiting lipid oxidation found in the ethanolic extracts of these plants was probably due to the greater removal of apolar antioxidants provided by alcohol compared to water. The low antioxidant activity in the extracts of aloe and *G. cambogia was probably due to* the low levels of carotenoids in the extracts of these plants.

To demonstrate the response of polar and apolar antioxidant groups in relation to the ABTS and p-carotene/linoleic acid methods, looking at Figure 1 (ABTS Test), it can be seen that quercetin and rutin, hydrophilic antioxidants, have a greater antioxidant potential than BHT (lipophilic antioxidant), i.e. in this test, the polar antioxidant groups showed better responses.

Table 2. Antioxidant activity of aqueous and ethanolic extracts of medicinal plants, herbal medicine and three standards, in % inhibition by the ß-carotene/linoleic acid method.

Plants		Water			Ethanol	
	40,000mg L "	20,000mg L "	10,000mg L "	40,000mg L "	20,000 mg L "	10,000 mg L "
Aloe vera	50.14±2.33 eA	39.44±0.37 eC	41.73 ± 0.09 dB	16.55±1.67 eD	12.27±2.09 eE	14.27±1.10 eE
Calunga	61.97±1.67 cB	58.05±0.85 bB	57.66±0.54 bB	70.75±0.17 dA	64.37±0.59 dA	46.67±3.41 cC
Carqueja	70.91±1.17 bA	60.21±1.32 bB	55.09±7.33 bB	73.53±1.37 cA	68.21 ±2.26 cA	49.67±0.29 cC
Garcinia cambogia	57.21±1.28 dA	54.13±3.08 cA	20.78 ± 0.28 dB	ND[1]	ND	ND
Marmelinho	99.98±1.10 aA	96.71 ±1.53 aA	86.08±7.44 aB	100.00±1.40 aA	100.00±0.34 aA	95.71±1.25 aA
Herbal medicine[2]	49.05±0.25 eD	48.41 ±1.50 dD	43.54±0.61 cE	87.23±0.34 bA	81.22±0.41 bB	56.58±4.15 bC
BHT (200 mg L ')			75.93±1.15			
Quercetin (200 mg L')[1]			70.48±0.71			
Rutin (200 mg L')[1]			8.71±1.29			

Data are the mean of three replicates ± standard deviation. Lowercase letters in the column compare between plants and uppercase letters in the row compare between concentrations. Equal letters do not differ according to the Scott-Knott test (P<0.05).[1] ND: Not detected. [2]Herbal medicine: made from a combination of Aloe vera, calunga, carqueja, Garcinia cambogia and marmalade in a ratio of 1:1.5:0.5:1.5:0.5 respectively.

Using the p-carotene/linoleic acid method (Table 2), BHT has the highest antioxidant potential, surpassing that of quercetin and rutin, i.e. non-polar antioxidants show better results.

When compared to the standards, the marmalade extracts were superior at all the concentrations tested, as was the ethanolic extract of the herbal medicine at concentrations of 40,000 and 20,000 mg L^{-1} , while the extracts of carqueja (aqueous and ethanolic) and calunga (ethanolic) at a concentration of 40,000 mg L^{-1} showed the same inhibition potential as quercetin. In relation to rutin, all the plants showed higher antioxidant potential than this standard. The different antioxidant activity results observed in the plant extracts are probably due to the different antioxidant compounds present in the extracts of these plants, and the differential principle employed by each antioxidant method used in this work. As a result, some plants showed good antioxidant potential for one method and poor potential for the other, highlighting the importance of using more than one technique to reflect the antioxidant capacity of a sample.

Studies carried out with these plants confirm this antioxidant potential. Rajasekaran et al. [35] observed that oral administration of the alcoholic extract of *Aloe vera* gel (300 mg kg^{-1}) to diabetic rats reversed the high levels of lipid peroxidation and hydroperoxides in the tissues of these rats to almost normal levels. The treatment also resulted in a significant increase in the levels of reduced glutathione, superoxide dismutase, catalase, glutathione peroxidase and glutathione-S-transferase in the liver and kidneys of the diabetic rats; these results show the antioxidant property of *Aloe vera* extract.

In vitro studies have shown that *Aloe vera* gel has the potential to capture free radicals [19,36], which was not observed in the present study, and the potential to combat the inhibition of lipid oxidation [36], corroborating the results of the present study. These differences may be due to various factors, such as the environment, the collection site, the age of the plant and the preparation and handling of the extracts. Yun et al. [37] analyzing *Aloe vera* at various ages, found that at the age of four this plant has a higher antioxidant potential than at younger ages, which shows that the stage of development plays a vital role in the antioxidant composition of *Aloe vera* gel.

For carqueja, Pádua et al. [39] reported a good antioxidant potential *in vitro* and *in vivo* for the hydroethanolic extract of this plant in a study carried out with neutrophils from Fisher rats. Mendes et al. [40] reported that the hydroalcoholic extract of carqueja has moderate antioxidant activity. In the same study, the authors observed reduced anti-ulcer activity in rats (stress ulcer induction). Dias et al. [41] and Morais et al. [2], analyzing the antioxidant potential of carqueja using the DPPH method, which has the same principle of capturing free radicals as the ABTS method, found a good antioxidant potential for carqueja extracts. These results differ from those found for carqueja in the present study, in which it was possible to observe a low antioxidant potential for the extracts (aqueous and ethanolic) using the ABTS method. This difference is probably due to factors such as where the plant was collected, its age and the forms and extractants used to prepare it.

For *G. cambogia*, Subhashini et al. [20] recorded a high antioxidant potential for the aqueous extract obtained from the peel of its fruit using the free radical scavenging (DPPH), reduced iron (FRAP), total reactive antioxidant potential (TRAP) and lipid peroxidation methods. These results differ from those found for the extracts (aqueous and ethanolic) of *G. cambogia* in the present study, when using the ABTS method, and show similar results when it comes to lipid oxidation. The different results between these studies may be due to the way in which the extracts were prepared, which is an extremely important factor in having good antioxidant activity. In this study, crude extracts were used, while in the study by Subhashini et al. [20] concentrated extracts were used, which may explain these differences. Other factors, such as geographical origin, the age of the plant and even adulteration and/or counterfeiting can be taken into account. In the literature consulted, no studies were found on the antioxidant potential of calunga and marmalade.

4. CONCLUSION

The plants studied showed antioxidant potential, and the antioxidant activity by the ABTS method was considered moderate in the marmalade extracts, and weak in the other plants. For the p-carotene/linoleic acid method, the aqueous and ethanolic extracts of marmalade, carqueja and calunga, and only the ethanolic extract of the herbal medicine showed good antioxidant potential.

These plants therefore have the potential to be used as antioxidant sources in pharmacological and food preparations, with possible health benefits.

CONSENT (IF APPLICABLE)

not applicable.

ETHICAL APPROVAL (WHERE APPLICABLE)

Not applicable.

ACKNOWLEDGMENTS

We would like to thank CAPES for the doctoral scholarship and FAPEMIG for the financial support and the scientific initiation scholarship.

AUTHORS' CONTRIBUTIONS

This work was carried out in collaboration between all the authors. Authors *AAS, ADC and TRM conceived* the study, authors FFL and PMBC carried out the experiments, and authors JMF and RMF managed the literature searches. Authors AAS, JMF and ADC were involved in writing the manuscript. All the authors read and approved the final manuscript.

COMPETING INTERESTS

The authors have declared that there are no competing interests.

ARTICLE 4

Effects of the administration of *Tournefortia paniculata* Cham. in *Wistar* Rats submitted to a hypercaloric diet: therapeutics and toxicology

Anderson Assaid Simão, Angelita Duarte Corrêa, Raimundo Vicente de Sousa, Vinicius de Oliveira Ramos, Silvana Marcussi

Submitted to Journal of Ethnopharmacology.

Article written in accordance with the journal's standards. This article is a preliminary version, considering that the journal's editorial board may suggest changes to adapt it to its own style.

Summary

Considering that medicinal plants are rich sources of bioactive compounds with potential for therapeutic use, widely used in folk medicine and in the preparation of industrialized medicines, the aim of this study was to evaluate the therapeutic and toxicological effects of administering the flour and aqueous extract of the leaves of *Tournefortia paniculata to Wistar* rats subjected to a hypercaloric diet for 42 days. The rats were divided into five groups and given the following treatments by gavage: T0 (control) - 1 mL of water day^{-1} ; T1 - aqueous extract containing 14 mg of phenolic compounds kg^{-1} rat day^{-1} ; T2 - 14 mg of quercetin kg^{-1} rat day^{-1} ; T3 - 50 mg of *T. paniculata* leaf meal (FTL) kg rat day^{-1-1} and T4 - 100 mg of FTL kg rat day^{-1-1} . The treatments did not significantly alter weight, but were effective in reducing liver fat, glucose and serum triglycerides. The T1 treatment reduced food consumption and lipid peroxidation. None of the treatments showed genotoxic potential. It can be concluded that *T. paniculata* leaves have anti-obesity potential. However, a more detailed study of the medicinal potential and the characterization of the phytochemicals present in this plant are still necessary for a better understanding of its mechanisms of action, enabling future applications in the treatment of this pathology or for various therapeutic purposes.

Keywords: obesity, medicinal plants, pharmacology, antioxidants, genotoxicity, herbal medicines.

1 - INTRODUCTION

Due to the consequences caused by obesity and the speed at which it spreads around the world, it has been considered a global epidemic, with more than one billion overweight adults, 400 million of whom are clinically obese, affecting both developed and developing countries, including Brazil (TUCCI et al., 2010). Its incidence is independent of socioeconomic factors and age, and its consequences range from the development of debilitating diseases to death, directly affecting individuals' quality of life.

Among the various causes of obesity, the most immediate determinant of excessive fat accumulation is the positive energy balance, which occurs when the amount of energy consumed is greater than the amount expended (HENRY; PANDIT, 2009). This is fully favored by current diets, which are high in energy density, rich in saturated fats, simple sugars and low in complex carbohydrates and fiber; it is also favored by the increase in physical inactivity, which shapes the contemporary Western lifestyle (POPKIN; GORDON-LARSEN, 2004) and contributes to the development of cardiovascular diseases, some types of cancer, diabetes, hypertension, among others, significantly affecting the national economy.

Among the options available for treating obesity, the most commonly used are balanced diets, regular exercise and drug treatments, ranging from lipase inhibitors to anorexics. However, due to the side effects and high cost of the drugs traditionally used to treat this disease, the potential of natural products for the treatment of obesity is being widely explored and could be a viable alternative for the future development of effective and safe drugs against obesity (MAYER et al., 2009; PARK et al., 2005).

A variety of natural products, including extracts and compounds isolated from plants, are being used to reduce body weight and prevent obesity (RAYALAM et al., 2008; SIMAO et al., 2012; SOUZA et al., 2012). In this context, plants that are popularly used for therapeutic purposes, which have no studies proving their effects, are of great value in research aimed at developing phytotherapics that can help in anti-obesity treatment and can act in its prevention or treatment, with proven efficacy and safety. Mixtures of phytochemicals or isolated molecules identified from these plants represent an excellent opportunity for the development of such therapeutics (BIRARI; BHUTANI, 2007).

Tournefortia paniculata Cham., traditionally known as marmalade, is a shrub belonging to the Boraginaceae family. Its origin is attributed to the Mediterranean regions and the United States, occurring widely in tropical and subtropical regions (MORAES; SOUSA, 2007). In Brazil, its leaves are used in folk medicine as a diuretic, anti-inflammatory in the urinary tract and in cases of nephrolithiasis (BERTOLUCCI, 2000).

Recent studies have reported that the aqueous extract of *T. paniculata* leaves may act in the treatment of obesity, due to its high *in vitro* inhibitory activity on the enzymes a-amylase and a-glucosidase (SIMAO et al., 2012). These enzymes are responsible for processing carbohydrates in the diet, acting in the breakdown of starch and the absorption of monosaccharides by enterocytes (OTA et al., 2009), making it a promising strategy to help treat obesity, hyperglycemia associated with type 2 diabetes and hypertension, due to the reduction in starch breakdown and glucose uptake in the intestine (KWON et al., 2006).

The antioxidant potential *in vitro* was also observed in the aqueous extract of these leaves (SIMAO et al., 2013) which, if demonstrated *in vivo,* will be characterized as an important tool to help treat obesity, since many diseases directly related to obesity, such as diabetes, cardiovascular diseases, hypertension, among others, can be caused or exacerbated by free radicals (VILLAREAL et al., 2005).

Simão et al. (2013) also found high levels of phenolic compounds in the leaves of *T. paniculata* (36.19 g 100 g^{-1} dry matter); these substances may have several benefits for the treatment of obesity and other diseases, since they have an inhibitory effect on digestive enzymes, as well as antioxidant and anti-glycemic properties, and phenolics may be the substances responsible for the possible anti-obesity effects of these leaves.

Despite being described as a constituent in the composition of the herbal medicine Moder Diet, revoked by the National Health Surveillance Agency and still available for purchase online, *T. paniculata* has no scientific studies to prove its pharmacological and/or toxic effects, and is not included in the official lists of medicinal plants released for the preparation of herbal medicines. Thus, its composition, rich in phenolic compounds, highlights this species as a strong candidate for drug development, although its extensive *in vitro* and *in vivo* pharmacotoxic characterization is still needed.

In this study, based on *in vitro* results, the potential therapeutic and toxicological effects of administering the flour (rich in phenolic compounds) and the aqueous extract of *T. paniculata* leaves on the metabolism of *Wistar* rats subjected to a hypercaloric diet were evaluated in order to determine possible anti-obesity properties, as well as safe and effective doses for the development of therapeutic formulations from this plant.

2 - MATERIAL AND METHODS

2.1 - Sample collection and preparation

The leaves of T paniculata Cham. (marmelinho) were purchased at the municipal market in Belo Horizonte, Minas Gerais, in January 2011 and transported to the laboratory. The leaves were washed in running distilled water and then placed in a forced-air oven to dry for 48 hours at ±35°C. After drying, the leaves were crushed in a Willey mill and then passed through an 80 mesh (0.2 mm) sieve, and only the plant material that passed through the sieve was used in the bioassay.

2.2 - Bioassay

2.2.1 - Animals and experimental conditions

The experiment was carried out in accordance with the ethical principles of animal experimentation, according to Law 11.794 of October 2008, and the project was approved by the Ethics Committee for the Use of Animals at the Federal University of Lavras (UFLA - Protocol 014/11).

The experiment was carried out over 10 weeks using 30 male *Wistar* rats (*Rattus norvegicus*) with an initial body weight of approximately 243.15 g. The animals were kept in individual cages at 21°C, with a light/ dark cycle. The animals were kept in individual cages at 21°C, 12-hour light/dark cycle, with access to distilled water *ad libitum*. During the first four weeks, the rats received a hypercaloric diet containing 46% commercial feed (Biobase Bio-tec Rats and Mice), 46% condensed milk and 8% corn oil (LEVIN et al., 1986). The diets were prepared weekly in 1 kg quantities and stored in a freezer at -24°C to minimize lipid oxidation of the polyunsaturated fatty acids. At the end of four weeks of consuming the hypercaloric diet, the animals were weighed, had a body weight of approximately 327.80 g, were divided into five groups of six animals each, and received the hypercaloric diet and the following treatments for six weeks: T0 (control) - 1 mL of water day^{-1} ; T1 - aqueous extract containing 14 mg of phenolic compounds kg^{-1} rat day^{-1} ; T2 (positive antioxidant control) - 14 mg of quercetin kg^{-1} rat day^{-1} ; T3 - 50 mg of flour from the leaves of *T. paniculata leaf meal* (FTL) kg rat day^{-1-1} and T4 - 100 mg FTL kg^{-1} rat day^{-1}

Each treatment was administered daily, diluted in water, and administered by gavage, using a sterilized stainless steel cannula and syringe, at the same time, by the same individual, with a volume of 1 mL per animal.

To calculate the dose for the administration of the aqueous extract and quercetin, the maximum dose of phenolics suggested for humans was taken into account, i.e. 1,000 mg per day^{-1} (SCALBERT et al., 2005), as well as the average weight of an adult human, which is 70 kg; in this way, it is possible to calculate the dose per gram of body weight, which is equivalent to 14 mg of phenolics per kg^{-1} of the animal's body weight. On the other hand, for the administration of FTL at a dose of 50 and 100 mg, the maximum solubility of this flour in water was taken into account.

To obtain the aqueous extract, the FTL was added to water in a ratio of 1:25 (w v^{-1}), kept stirring for 60 minutes at room temperature and then centrifuged at 10,000 x g for 10 minutes. The supernatant was collected and the phenolic compounds measured using the Folin-Denis reagent, with tannic acid as the standard (AOAC, 2005). The phenolic compound levels found in this extract were 16.25 mg mL^{-1} and in the FTL 40.62 mg 100 mg^{-1} .

On the other hand, for quercetin and FTL, aqueous solutions were prepared with concentrations of 20 and 40 mg mL^{-1} , respectively. The volume administered in each treatment was calculated according to the weight of each rat. Thus, a volume of solution corresponding to the dose calculated according to the body weight measured on the day was administered to each animal by gavage. The same was done for the control group, which was given only water.

2.2.2 - Body weight and daily food consumption

Body weight and food consumption were recorded daily to assess the parameters weight gain and food consumption, with food consumption calculated as the difference between the diet offered and the leftovers.

2.2.3 - End of bioassay

At the end of the experiment, blood samples were taken from the tail end, after a 12-hour fast, to determine glucose concentrations, using a portable glucometer (Accu-Chek® - Active - São Paulo,

Brazil) and reagent strips (Accu-Chek Advantage II).

The rats were then anesthetized with sodium thiopental (1 g of sterile powder dissolved in 40 mL of 0.9% saline solution), intraperitoneally, at a dose of 40 mg kg^{-1} . After collecting 2 to 4 mL of blood by cardiac puncture, which caused death by exsanguination, an aliquot of 50 ML was separated for genotoxicity assessment (comet assay), and the rest of the blood was centrifuged at 2,500 x g for 10 minutes to obtain serum, which was kept under refrigeration.

After euthanasia, a necropsy was also carried out, removing the internal organs (heart, liver, kidneys and pancreas), weighing the liver and preparing histological slides. The liver was divided into three parts: one for qualitative analysis of the histology of this organ, one for determination of lipid peroxidation and one for determination of the ether extract. After removing the organs, the head, legs, tail and skin were removed, followed by washing the carcass with a saline solution.

2.2.4 - Liver and carcass lipid analysis

The carcass and part of the liver were weighed and freeze-dried until constant weight. They were then ground in a mortar and lipids were measured using the Soxhlet method (AOAC, 2005).

2.2.5 - Laboratory blood tests

The blood serum was used to determine the concentrations of glucose, total cholesterol, HDL cholesterol, triglycerides, the activity of the enzymes aspartate aminotransferase (AST), alanine aminotransferase (ALT) and gamma glutamyl transferase (GGT). The blood constituents were analyzed using commercial kits (Labtest) and the analyses were carried out using the Automatic Veterinary Biochemical Analyzer Thermoplate.

2.2.6 - Lipid peroxidation

Lipid peroxidation was determined by the formation of thiobarbituric acid reactive substances (TBARS), according to Winterbourn et al. (1985).

The livers of the animals were weighed and homogenized in a Potter tissue homogenizer, in an ice bath, after adding 0.1 mol L sodium phosphate buffered $saline^{-1}$ (PBS), pH 7.4 (volume equivalent to 4 times the fresh weight of the tissue). The homogenate was centrifuged at 3,000 x g for 10 minutes at 4°C, and the supernatant, kept in an ice bath, was used in the tests. Aliquots of 500 ^L of the supernatant were mixed with 500 |1L of 25% hydrochloric acid (v v^{-1}), 500 LII. of 1% thiobarbituric acid (w v^{-1} , in 0.05 mol L^{-1} NaOH) and 45 |iL of 2% BHT (w v^{-1} , in ethanol). The mixture was heated in a boiling water bath for 10 minutes. After cooling in an ice bath, 1.5 mL of butanol was added and the samples were shaken vigorously, centrifuged at 900 x g for 5 minutes and the fraction containing butanol was collected and used to determine the absorbance at 532 nm.

The concentration of TBARS was calculated from the standard curve of malondialdehyde - MDA (1,1,3,3 tetraethoxypropane). The results were expressed in moles MDA mg^{-1} fresh sample.

2.2.7 - Comet test

2.2.7.1 - Obtaining nucleoids and electrophoresis

The comet assay was carried out according to the methodology described by Singh et al. (1988), with some modifications. An aliquot (15 uL) of each cell suspension containing the treatments was mixed with 100 p.L of low melting point agarose (0.5% w v^{-1} in PBS), applied to a microscope slide previously coated with normal melting point agarose solution (1% w v^{-1} in PBS), covered with a coverslip (24 x 60 mm) and kept at ± 4°C for 5 minutes, until the agarose solidified. For each treatment/rat, 2 slides were prepared. The coverslips were then removed and the slides were immersed in a lysis solution (2.5 mol l^{-1} NaCl, 100 mmol l^{-1} EDTA, 10 mmol l^{-1} Tris, 1% Triton X-100, 10% DMSO; pH 10), where they remained for 24 hours at 4°C.

After lysis, the slides were kept for 25 minutes in a freshly prepared electrophoresis solution (1 mmol L^{-1} EDTA, 300 mmol L^{-1} NaOH; pH 13), allowing the bonds in the DNA molecules to loosen and the labile sites to be exposed to alkalis. Electrophoresis was carried out at 25V for 35 minutes. After electrophoresis, the slides were kept in a neutralization solution (0.4 mol L^{-1} Tris; pH 7.5) for 30 minutes, then dried and fixed with 100% ethanol. All procedures were carried out in the dark.

2.2.7.2 - Staining and analysis

The slides were stained with 45 ul. of propidium iodide solution (1mg mL^{-1}), covered with a coverslip and analyzed under an epifluorescence microscope (Nikon EClIPSE E400) at 200x magnification.

To measure the levels of damage to the DNA molecules of the lymphocytes, 100 nucleoids from each slide, 2 slides per treatment/rat (totaling 200 nucleoids per treatment/rat), were counted and classified by the same evaluator, using visual scoring standards described by Singh et al. (1988). The cells were classified according to the size of the "tail" and the diameter of the "head" into Class 0: no damage (damage <5%); Class 1: low level of damage (5-20%); Class 2: intermediate level of damage (20-40%); Class 3: high damage (40-85%); Class 4: totally damaged (damage > 85%). The average frequency of damage was calculated from the sum of the percentages of damage 1, 2, 3 and 4. The arbitrary units (0-400; where 0 = no damage and 400 = 100% damage) were calculated using the equation (1 x number of nucleoids grouped in class 1) + (2 x number of nucleoids in class 2) + (3 x number of nucleoids in class 3) + (4 x number of nucleoids in class 4), as described by Collins (2004).

2.2.8 - Histopathology

For the histopathological analysis, slides were prepared containing sections of the heart, liver, kidneys and pancreas of each rat. For histological processing, the organs were cut into small pieces, immersed in 10% formalin for 24 hours and kept in 70% alcohol until they were embedded in paraffin. 4-Lim sections were cut on a rotary microtome, stained with hematoxylin-eosin, mounted on slides and analyzed under light microscopy (100x magnification) for qualitative analysis.

2.3 - Statistical analysis

The experiment was conducted in a completely randomized design. For the weight gain data, the plots were subdivided into a 5 x 7 scheme, i.e. five treatments, a seven-week experiment with six repetitions, and for the feed consumption data, the plots were subdivided into a 5 x 6 scheme, with five treatments, a six-week experiment with six repetitions. For the other analyses, five treatments with six replications were used and, when the analysis of variance showed significance, the means were compared using the Scott-Knott test (P<0.05), using the Sisvar program (FERREIRA, 2003).

3 - RESULTS AND DISCUSSION

Figure 1 shows the body weight and food consumption of the animals submitted to the different treatments during the six weeks of the experiment. The average weights of the rats did not differ significantly between treatments (Figure 1). However, the average weight gain of the animals in the T0, T2, T3 and T4 groups was 10.20%, 4.57%, 3.07% and 7.11%, respectively, while for the T1 group, the

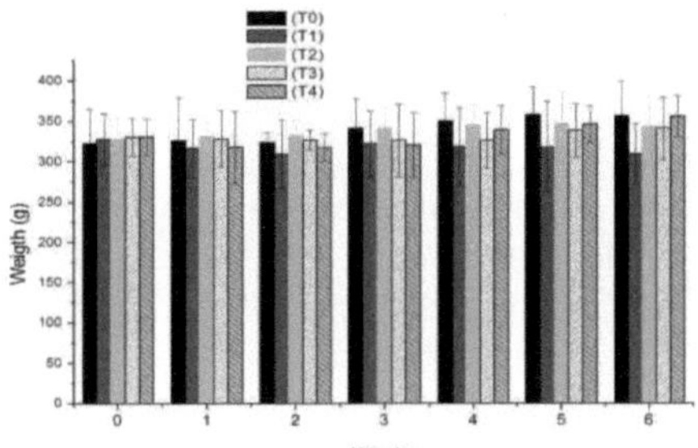

Figure 1. Body weight of *Wistar* rats subjected to treatments: T0 (control) - 1 mL of water day^{-1} ; T1 - aqueous extract containing 14 mg of phenolic compounds kg^{-1} rat day^{-1} ; T2 - 14 mg of quercetin kg^{-1} rat day^{-1} ; T3 - 50 mg of *T. paniculata* leaf meal (FTL) kg^{-1} rat day^{-1} and T4 - 100 mg of FTL kg rat day^{-1-1} , for six weeks. The data represent the mean of six repetitions ± standard deviation of the mean weight for each week of the experiment.

Weight loss was observed in the animals, with an average weight gain of 5.80%, although this was not statistically different.

In another study with the aqueous extract of *T. paniculata* leaves, an *in vitro* inhibition of the enzymes a-amylase and a-glucosidase was observed, with this inhibitory effect being one of the possible mechanisms responsible for the weight reduction, considering a lower absorption of carbohydrates (SIMAO et al., 2012). Souza et al. (2012) evaluated the anti-obesity effects of the methanolic extract of *Baccharis trimera* leaves, administered to *Wistar* rats by gavage, and also observed a reduction in the animals' weights and suggested a possible mechanism of action based on the inhibition of lipases and glycosidases, which were inhibited *in vitro* by the methanolic extract of this plant. These reports show that medicinal plants with digestive enzyme inhibitors in their extracts, such as *T. paniculata,* can be useful for limiting the absorption of fats and carbohydrates from the diet, resulting in weight loss.

Some studies have described quercetin's anti-obesity potential through mechanisms such as increasing lipolysis, inhibiting adipogenesis and inducing apoptosis in adipocytes (HSU; YEN, 2006; AHN et al., 2008; FANG, 2008). However, in this study, quercetin (T2) was not able to cause a reduction in body weight, confirming the results of Machha et al. (2007), who also found no reduction in the body weight of *Wistar* rats after treatment with quercetin, administered by gavage, at a dose of 10 mg kg^{-1} rat for six weeks.

The average daily feed consumption by the animals is shown in Figure 2. Treatment T1 had a significant effect, reducing food consumption.

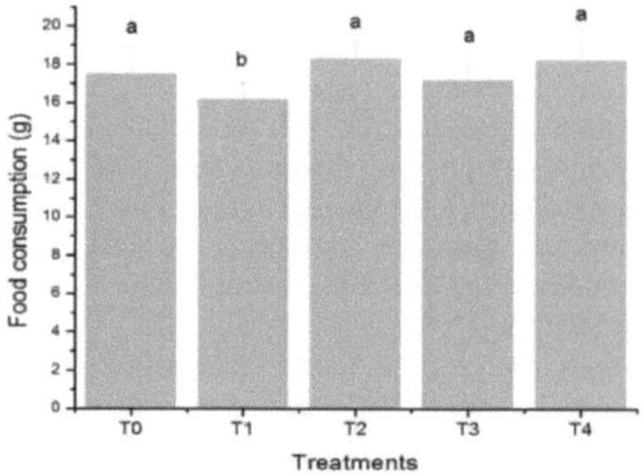

Figure 2. Food consumption of *Wistar* rats subjected to the treatments: T0 (control) - 1 mL of water day^{-1} ; T1 - aqueous extract containing 14 mg of phenolic compounds kg^{-1} rat day^{-1} ; T2 - 14 mg of quercetin kg^{-1} rat day^{-1} ; T3 - 50 mg of *T. paniculata* leaf meal (FTL) kg^{-1} rat day^{-1} and T4 - 100 mg of FTL kg rat day^{-1-1} , for six weeks. The same letters in the columns do not differ according to the Scott-Knott test (P <0.05).

These results showed that the T1 treatment was effective in reducing appetite, which characterizes an anorexigenic action by an unknown mechanism, and this inhibition may be related to the weight loss observed for this treatment (Figure 1).

Table 1 shows the average liver weights, the liver weight/carcass weight ratio and the body and liver fat levels, without

Table 1. Average liver weight, liver weight/carcass weight, body and liver fat of *Wistar* rats submitted to[1] treatments for six weeks.

Parameter	TO	T1	T2	T3	T4
Liver weight (g)	7.71±1.10^{a}	7.12±1.39^{a}	7.79±1.01^{a}	7.95±1.36^{a}	8.24±0.74^{a}
Liver weight/total body weight (%)	4.49±0.85^{a}	4.54±0.70^{a}	4.69±0.27^{a}	4.82±0.60^{a}	4.99±0.40^{a}
Liver fat (g 100 g')[1]	13.63±0.84^{a}	11.13±1.39^{b}	10.63±1.05^{b}	8.27±0.99^{c}	8.89±1.25^{c}
Body fat (g 100 g')[1]	18.92±1.93^{a}	17.24±1.61^{a}	17.76±2.74^{a}	17.01±1.98^{a}	19.35±1.10^{a}

Data are the mean of six replicates ± standard deviation, treatments: TO (control) - 1 mL of water day'[1] ; T1 - aqueous extract containing 14 mg of phenolic compounds kg'[1] rat day'[1] ; T2 - 14 mg of quercetin kg'[1] rat day'[1] ; T3 - 50 mg of *T. paniculata* leaf meal (FTL) kg'[1] rat day'[1] and T4 - 100 mg of FTL kg' rat day'[11] . The same letters in the lines do not differ according to the Scott-Knott test (P <0.05).

Statistical difference between treatments for average liver weights, liver weight/carcass weight ratio and body fat levels. The treated groups had significantly lower liver fat levels than the T0 (control) group, with a reduction of 18.34%, 22.01%, 39.32% and 34.77% for the T1, T2, T3 and T4 groups, respectively.

Considering that the liver would be affected by possible anti-nutritional factors present in *T. paniculata* leaves, the damage could reflect in anatomical alterations in the treated groups, which did not occur, thus ruling out the induction of hypertrophy and hyperplasia under the conditions of this experiment.

Body fat has been used as an indicator of obesity in rats. The accumulation of adipose tissue has been associated with various markers of obesity, such as weight gain, body mass and the Lee index (NASCIMENTO et al., 2008).

Table 2 presents the results of some biochemical indicators measured in the blood of the animals under study.

Determining glucose is the initial parameter for detecting alterations and disorders in glucose metabolism, such as type 2 diabetes, which is generally associated with obesity (MOTA, 2009).

Table 2. Biochemical parameters of *Wistar* rats submitted to[1] treatments for six weeks.

Parameters	TO	T1	T2	T3	T4
Caudal capillary glucose	97.16±2.48^{a}	82.5±5.50^{b}	84.16±5.38^{b}	86.83±7.83^{b}	93.66±3.44^{a}
(mg dL')[1] Serum glucose (mg dL')[1]	165.16±17.17^{a}	140.34±6.12^{b}	168.08±9.34^{a}	162.31±8.09^{a}	177.96±6.35^{a}
Total cholesterol (mg dL')[1]	56.16±4.11^{a}	55.19±5.36^{a}	61.48±6.67^{a}	59.63±4.26^{a}	58.13±4.56^{a}
HDL cholesterol (mg dL')[11]	49.11±5.36^{a}	43.77±2.38^{a}	47.64±1.27^{a}	45.11±3.91^{a}	44.33±4.12^{a}
Triglycerides (mg dL')[1]	42.51±1.92^{a}	29.23±4.00^{b}	31.70±8.73^{b}	34.40±3.68^{b}	32.80±2.86^{b}
AST[2] (U L')[1]	77.35±0.57^{a}	80.45±4.06^{a}	74.35±5.14^{a}	71.58±5.88^{a}	76.12±4.82^{a}

ALT[3] (U L')[1]	59.74±6.32[b]	87.25±16.14[a]	63.17±7.35[b]	56.82±3.15[b]	61.49±4.85[b]
GGT[4] (U L)[1]	2.55±0.00[a]	2.48±0.07[a]	2.48±0.07[a]	2.42±0.19[a]	2.50±0.06[a]

Data are the mean of six replicates ± standard deviation, treatments: TO (control) - 1 mL of water day'[1] ; T1 - aqueous extract containing 14 mg of phenolic compounds kg'[1] rat day'[1] ; T2 - 14 mg of quercetin kg' rat day'[11] ; T3 - 50 mg of *T. paniculata* leaf meal (FTL) kg'[1] rat day'[1] and T4 - 100 mg of FTL kg' rat day'[11] . [2]AST: aspartate aminotransferase. [3]ALT: alanine aminotransferase. [4]GGT: gamma glutamyl transferase. Equal letters in the lines do not differ according to the Scott-Knott test (P <0.05).

The levels of capillary glucose in the animals subjected to treatments T1, T2 and T3 were significantly lower than in the T0 (control) group, with a 15.09%, 13.38% and 10.63% reduction in glucose concentration, respectively.

For serum glucose, only treatment T1 was able to significantly reduce the concentration of glucose compared to group T0 (control), demonstrating the greater hypoglycemic potential of the aqueous extract of *T. paniculata* leaves compared to the other treatments. This reduction may be due to the inhibition of the enzymes a-amylase and a-glucosidase, reducing the concentration of glucose through the lower absorption of carbohydrates (SIMAO et al., 2012), and through the lower food consumption observed for this treatment (Figure 2).

Some authors have attributed the hypoglycemic effect of plants to the phenolic compounds present in their extracts (FIGUEROA- VALVERDE et al., 2009; URZEDA et al., 2013), and these compounds, present in the leaves of *T. paniculata,* may be responsible for the reduction in glucose concentration observed in this study.

Blood levels of lipids and liver enzymes provide important parameters in determining the safety of functional ingredients or plant-derived end products subjected to toxicity tests (PATEL et al., 2008).

The results showed that there was no statistical difference between the groups in terms of the serum concentration of total cholesterol and HDL cholesterol. The treated groups did not differ statistically, and triglyceride concentrations were significantly lower than the T0 (control) group, with a 31.24%, 25.43%, 19.08% and 22.85% reduction in concentration for the T1, T2, T3 and T4 groups, respectively.

Studies have shown a significant reduction in the concentration of glucose in the blood of diabetic rats treated with quercetin (COSKUN et al., 2005; MACHHA et al., 2007), as well as a reduction in triglycerides (RICARDO et al., 2001), and these studies are consistent with the results of the present study, in which a reduction in the concentration of glucose and triglycerides was observed in the rats treated with quercetin (T2).

Based on the data in Table 2, the AST and GGT activities were statistically the same for all the groups. With regard to ALT, the T1 treatment showed significantly higher activity than the other groups, with a 46.04% increase in activity compared to the T0 (control) group. Thus, the ALT results for the T1 treatment suggest that this treatment may cause hepatotoxic effects.

The increase in serum levels of ALT enzyme activity caused by T1 treatment can probably be attributed to liver dysfunction due to hepatocyte rupture, resulting from necrosis or changes in cell membrane permeability (KANEKO, 1989). Injury or destruction of liver cells releases transaminases into the bloodstream (MOTA, 2009), resulting in an increase in their activity.

AST, which is mainly found in the mitochondria (80%), in organs such as the heart, liver, kidneys, pancreas and skeletal muscles, is not released as quickly as ALT, which is essentially hepatic and located in the cytoplasm, thus justifying the fact that in this study only statistical changes were recorded for ALT concentrations in relation to the control group. Furthermore, ALT is a more sensitive indicator of acute hepatotoxicity than AST (AL-HABORI et al., 2002).

The oxidative stress index, normally caused by free radicals in the body, was determined by lipid peroxidation, which is considered an important marker of oxidative stress and also one of the main factors involved in cell damage caused by these radicals (BALU et al., 2005).

The lipid peroxidation results (Figure 3) for the T1 and T2 treatments were significantly lower than those of the T0 (control) group, with a reduction of 34.09% and 18.90%, respectively, in the MDA concentration.

Quercetin (T2) was used as a positive control, as it is a flavonoid with proven antioxidant activity (DUARTE-ALMEIDA et al., 2006; SIMAO et al., 2013). Through *in vivo* studies, it was observed that quercetin was effective in reducing TBARS in rat livers (GARCiA-SAURA et al., 2005; YAMAMOTO; OUE, 2006), justifying its use as a reference substance. It has five reactive phenolic hydroxyls and high resonance stability, these properties being responsible for its antioxidant potential and possibly for the reduction in MDA concentration seen in this study.

On the other hand, treatment T1, which contains phenolic compounds, may contribute, at least in part, to the antioxidant activity demonstrated in this study, since these compounds have redox properties that allow them to act as hydrogen-donating reducing agents. However, the complexation of Fe^{+2} with phenolic compounds may reduce the availability of this metal involved in the Fenton reaction, initiation and

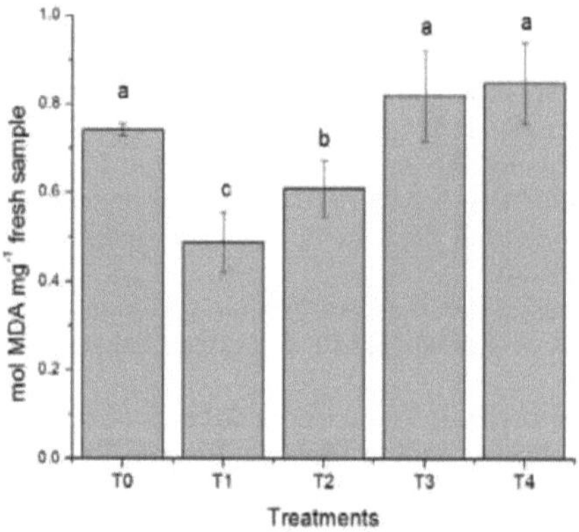

Figure 3: Production of thiobarbituric acid reactive substances in the liver tissue of *Wistar* rats subjected to the treatments: T0 (control) - 1 mL of water day^{-1} ; T1 - aqueous extract containing 14 mg of phenolic compounds kg^{-1} rat day^{-1} ; T2 - 14 mg of quercetin kg^{-1} rat day^{-1} ; T3 - 50 mg of *T. paniculata* leaf meal (FTL) kg^{-1} rat day^{-1} and T4 - 100 mg of FTL kg rat day^{-1-1} , for six weeks. [1]MDA: malondialdehyde. Data are the mean of six replicates ± standard deviation. The same letters in the columns do not differ by the Scott-Knott test (P <0.05).

propagation of lipid peroxidation (LIMA et al., 2006), which could explain the reduction in MDA concentration by the aqueous extract of *T. paniculata* leaves. For this study, treatment T1 showed better results for antioxidants than the phenolic standard of quercetin, demonstrating the good antioxidant potential of this plant.

With regard to treatments T3 (50 mg of FTL kg^{-1} rat day^{-1}) and T4 (100 mg of FTL kg^{-1} rat day^{-1}), containing 20.31 and 40.62 mg of phenolics respectively, neither was able to inhibit the formation of lipid peroxidation products, suggesting that the antioxidant substances present in this flour are only able to exert antioxidant effects when they are extracted and concentrated.

It is worth noting that the treatments used in this study, with the exception of treatment T2 (quercetin), were not subjected to any purification process to obtain the active ingredients, i.e. only the flour and crude aqueous extract of *T. paniculata* leaves were used, showing its therapeutic potential as an antioxidant and hypoglycemic, even when used without any purification process.

The results relating to the fragmentation of DNA molecules (Table 3) showed a predominance of nucleoids with no damage (class 0), with average percentages ranging from 43.16% (T1) to 83.05% (T0) and with a low level of damage (class 1), with averages ranging from 16.32% (T0) to 52.58% (T1). It was also possible to observe (Table 3) a reduced number of nucleoids with an intermediate level of damage (class 2), with average percentages ranging from 0.00% (T4) to 5.29% (T1) and an absence of nucleoids with a high level of damage (class 3) and totally damaged DNA (class 4) in all the treatments.

Table 3. Average number of nucleoids per comet class, frequency of nucleoids with damage and arbitrary units in leukocytes from *Wistar* rats submitted to treatment for six weeks.

Treatments[2]	Comet class (%)*					Frequency of damage[3] (%)	Arbitrary units[4]
	0	1	2	3	4		
TO	83.05±4.28^{a}	16.32±3.68^{d}	0.98±0.08^{b}	0.00±0.00^{a}	0.00±0.00^{a}	17.30±3.87^{d}	18.28±4.20^{d}
T1	43.16±16.53^{d}	52.58±11.71^{a}	5.29±1.18^{a}	0.00±0.00^{a}	0.00±0.00^{a}	57.87±11.86^{a}	63.16±13.56^{a}
T2	71.30±5.50^{b}	28.40±5.11^{c}	0.30±0.03^{b}	0.00±0.00^{a}	0.00±0.00^{a}	28.70±5.50^{c}	29.00±5.89^{c}
T3	59.60±5.86^{c}	39.30±4.66^{b}	1.13±0.24^{b}	0.00±0.00^{a}	0.00±0.00^{a}	40.76±5.59^{b}	41.56±5.95^{b}
T4	68.18±9.63^{b}	31.83±9.63^{c}	0.00±0.00^{b}	0.00±0.00^{a}	0.00±0.00^{a}	31.83±9.63^{c}	31.83±9.63^{c}

The data represent the average values obtained in 2 slides, with 100 nucleoids analyzed in each slide, and 4 slides/treatment/rat. 30 rats were used (6 rats/treatment). treatments: TO (control) - 1 mL of water day'1 ; T1 - aqueous extract containing 14 mg of phenolic compounds kg'1 rat day'1 ; T2 - 14 mg of quercetin kg'1 rat day'1 ; T3 - 50 mg of flour from the leaves of *T. paniculata leaf meal* (FTL) kg'1 rat day'1 and T4 - 100 mg FTL kg'1 rat day'1 , damage frequency: sum of damage classes 1 to 4;[4] Arbitrary units: (1 x damage class 1) + (2 x damage class 2) + (3 x damage class 3) + (4 x damage class 4). Equal letters in the columns do not differ according to the Scott-Knott test ($P < 0.05$).

Although the percentages of nucleoids with damage and arbitrary units show significant differences between the T0 group (control) and the other treatments, these values are related to damage in classes 1 (low level of damage) and 2 (intermediate damage), demonstrating the absence of genotoxicity, considering the six weeks of exposure to the treatments, thus suggesting that, in the composition of the treatments evaluated, there are no molecules capable of causing genotoxic and, consequently, mutagenic effects on DNA under the experimental conditions used.

Some studies have used the comet assay to complement the toxicological characterization of plant extracts, investigating the cytotoxic and genotoxic effects of secondary metabolites (phenolic compounds and terpenes), such as those carried out by Pereira et al. (2012), who investigated the genotoxic potential of white bean flour, containing tannins and other phenolic compounds, on human leukocytes, describing the prevalence of low levels of damage in their results; that carried out by Wan-Ibrahim et al. (2010), who evaluated 20 aqueous plant extracts and observed that only two of them, e.g. *Vitex Pinnata* L. and *Quercus infectoria* Oliver, caused more than 50% damage to the DNA of human leukocytes, which is attributed to their phenolic content, specifically tannic acid; and those of Strange et al. (2009), who analyzed natural products such as guaco (*Mikania glomerata* Spreng), espinheira-santa (*Maytenus ilicifolia* Mart. ex Reiss) and sage (*Lippia alba* Mill. NE) and observed cytotoxic and genotoxic activities induced by metabolites such as coumarins, tannins and terpenes present in the extracts of these plants, highlighting the need for a toxicological characterization of plant metabolites, since they can induce side and/or therapeutic effects depending on factors such as dose and chronic use.

Figure 4 shows images of histological sections taken from different organs. No macroscopic alterations were observed in the organs analyzed, nor were there any significant lesions or microscopic alterations. The possible accumulation of fat in the liver of the animals induced by the hypercaloric diet was not observed in the histology, nor was the damage or destruction of liver cells that could release transaminases into the bloodstream, causing an increase in their activity. However, the length of time of the experiment may not have been sufficient to reveal such damage under light microscopy, since there was an increase in ALT activity for the T2 treatment (Table 2) and for the

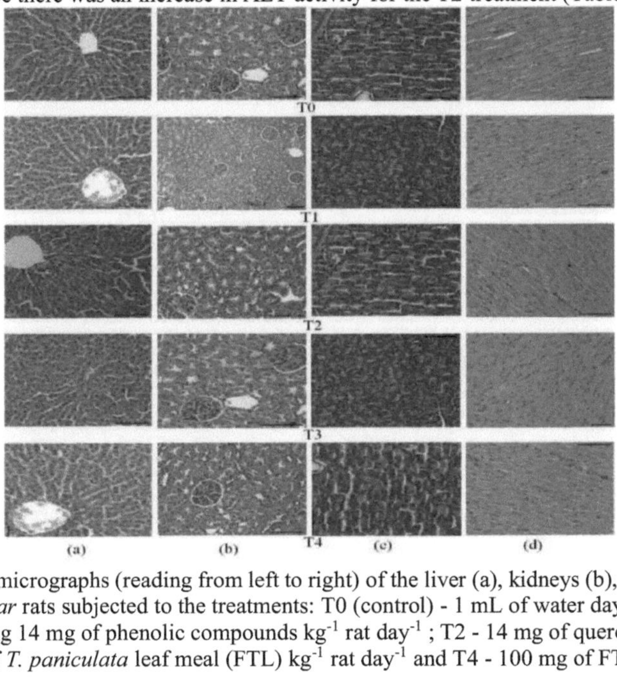

Figure 4: Photomicrographs (reading from left to right) of the liver (a), kidneys (b), pancreas (c) and heart (d) of *Wistar* rats subjected to the treatments: T0 (control) - 1 mL of water day^{-1} ; T1 - aqueous extract containing 14 mg of phenolic compounds kg^{-1} rat day^{-1} ; T2 - 14 mg of quercetin kg^{-1} rat day^{-1} ; T3 - 50 mg of *T. paniculata* leaf meal (FTL) kg^{-1} rat day^{-1} and T4 - 100 mg of FTL kg rat day^{-1-1} ,

for six weeks.
a decrease in the concentration of liver fat compared to the T0 (control) group (Table 1).

The results obtained for the leaves of *T. paniculata* in this study showed the presence of anti-obesity and antioxidant properties, as well as the absence of genotoxicity, highlighting the high pharmaceutical potential of this plant, which can be widely studied as a source of biomolecules or molecular models for the development of new pharmaceutical and/or cosmetic formulations. These results show prospects for future studies to determine doses and investigate other parameters that define efficacy and safety.

4 - CONCLUSION

The leaves of *T. paniculata* have anti-obesity potential, as they cause a reduction in serum glucose and triglyceride levels, as well as in liver fat (flour and water extract) and food consumption, and also in lipid peroxidation (aqueous extract), in addition to not causing genotoxic effects under the conditions evaluated. However, more detailed studies on the medicinal and toxicological potential, as well as the characterization of the phytochemicals present in this plant, are still necessary for a better understanding of its mechanisms in the treatment of this pathology, in addition to other possible therapeutic applications.

ACKNOWLEDGMENTS

The authors would like to thank CAPES for the doctoral scholarship and FAPEMIG for the financial support.

REFERENCES

NATIONAL HEALTH SURVEILLANCE AGENCY. **Resolution RE No. 3.691**, of February 15, 2009. Prohibits the manipulation and commercialization, throughout the national territory, of the herbal medicine Moder Diet, as a safety measure for not meeting the regulatory requirements of this agency. Brasília, 2009. Available at: <http. www.anvisa.org.br>. Accessed on: September 20, 2009.

AGOSTINI, F. et al. Study of the essential oil of some species of the genus *Baccharis* (Asteraceae) from southern Brazil. **Revista Brasileira de Farmacologia**, São Paulo, v. 15, n. 3, p. 215-220, jul./set. 2005.

ANTUNES, A. F. **Inhibitory activity of cerrado plant extracts on alpha-amylases**. 2008. 96 p. Dissertation (Master's Degree in Health Sciences) - University of Brasília, Brasília, 2008.

BRAZILIAN ASSOCIATION FOR THE STUDY OF OBESITY AND METABOLIC SYNDROME. **Etiology of obesity**. São Paulo, 2009. Available at: <http://www.abeso.org.br/pdf/Etiologia%20e%20Fisiopatologia%20-%20Walmir%20Coutinho.pdf>. Accessed on: September 15, 2010.

BACH, D. B.; LOPES, M. A. Estudo da viabilidade econômica da cultura de *Aloe vera* L. **Ciência e Agrotecnologia**, Lavras, v. 31, n. 4, p. 1136-1144, jul./ago. 2007.

BAGGIO, C. H. et al. Gastroprotective effects of a crude extract of *Baccharis illinita* DC in rats. **Pharmacollogy Research**, Bressanone, v. 47, n. 1, p. 93-98, jan. 2003.

BARBOSA-FILHO, J. M. et al. Plants and their active constituents from South, Central and North America with hypoglycemic activity. **Revista Brasileira de Farmacognosia**, São Paulo, v. 15, n. 4, p. 392-413, Oct./Dec. 2005.

BARCELOS, M. F. P. **Natural toxic substances in food**. Lavras: UFLA, 2004. 114 p.

BARREIROS, A. L. B. S.; DAVID, J. M.; DAVID, J. P. Estresse oxidativo: relação entre geração de espécies reativas e defesa do organismo. **Química Nova**, São Paulo, v. 29, n. 1, p. 113-123, feb. 2006. BERTOLUCCI, S. K. V. et al. Micropropagation of *Tournefortiapaniculata* Cham. **Revista Brasileira de Plantas Medicinais**, Botucatu, v. 3, n. 1, p. 43-49, 2000.

BHUTANI, K. K.; BIRARI, R.; KAPAT, K. Potential anti-obesity and lipid-lowering natural products: a review. **Natural Product Communications**, Westerville, v. 2, n. 1, p. 331-348, 2007.

BIANCHI, M. L. P.; ANTUNES, L. M. G. Free radicals and the main antioxidants in the diet. **Revista de Nutri^ão**, Campinas, v. 12, n. 2, p. 123-130, May/Aug. 1999.

BIRARI, R. B.; BHUTANI, K. K. Pancreatic lipase inhibitors from natural sources: unexplored potential. **Drug Discovery Today**, London, v. 12, n. 19/20, p. 879-889, Oct. 2007.

BLUMENKRANTZ, M. **Obesity:** the world's metabolic disorder. Beverly Hills: Quantumhcp, 1997. Available at: <http://www.quantumhcp.com/obesity.htm>. Accessed on: March 10, 2013.

BOUDREAU, M. D.; BELAND, F. A. An evaluation of the biological and toxicological properties of *Aloe barbadensis* (Miller), *Aloe vera*. **Journal of Environmental Science and Health Part C - Environmental Carcinogenesis & Ecotoxicology Reviews**, Philadelphia, v. 24, n. 1, p. 103-154, Jan./June 2006.

BRAZIL. **Normative Instruction No. 5**, of December 11, 2008. Determines the publication of the list of herbal medicines for simplified registration. Brasília, 2008. Available at: <http://www.portal.saude.gov.br/portal/arquivos/pdf/IN_N_5_2008_anvisa.pdf> . Accessed on: Dec. 10, 2012.

BRAZIL. **RDC No. 10**, of March 10, 2010. Provides for the notification of plant drugs to the National Health Surveillance Agency (ANVISA) and other measures. Brasília, 2010. Available at: <http://www.brasilsus.com.br/legislacoes/rdc/103202-10.html>. Accessed on: Dec. 10, 2012.

BRAZIL. Ministry of Health. **Primary care books:** obesity. Brasília, 2006. Available at: <http://portal.saude.gov.br/portal/saude/default.cfm>. Accessed on: September 15, 2010.

BRAZIL. Ministry of Health. **National list of plants of interest to SUS (RENISUS)**. Brasília, 2009. Available at: <http://portal.saude.gov.br/portal/arquivos/pdf/RENISUS_2010.pdf>. Accessed on: May 10, 2013.

BRAZIL. Ministry of Health. **RENAME**. Brasília, 2012a. Available at:

<http://portal.saude.gov.br/portal/arquivos/pdf/anexos_rename_2012_pt_533_11 _06_2012.pdf>. Accessed on: May 2, 2012.
BRAZIL. Ministry of Health. Health Surveillance Secretariat. **Surveillance of risk and protective factors for chronic diseases by telephone survey**. Brasília, 2012b. Available at: <http://portal.saude.gov.br/portalsaude/arquivos/pdf/2012/Ago/22/vigitel_2011_ final_0812.pdf>. Accessed on: December 1, 2012.
CAPASSO, R. et al. Phytotherapy and the quality of herbal medicines. **Fitoterapia**, Amsterdam, v. 71, n. 1, p. 58-63, Aug. 2000.
CARVALHO, A. C. B. et al. Status of the registration of herbal medicines in Brazil. **Revista Brasileira de Farmacognosia**, Joao Pessoa, v. 18, n. 2, p. 314-319, Apr./Jun. 2008.
CENTERS FOR DISEASE CONTROL AND PREVENTION. **Study estimates that the medical cost of obesity could reach US$ 147 billion annually**. Available at: <http://www.cdc.gov/media/pressrel/2009/r090727.htm>. Accessed on: Dec. 10, 2012.
CHOKSHI, D. Subchronic oral toxicity of standardized extract of white kidney bean (*Phaseolus vulgaris*) in rats. **Food and Chemical Toxicology**, Oxford, v. 45, n. 1, p. 32-40, Jan. 2007.
DANIELS, S. R. Complications of obesity in children and adolescents. **International Journal of Obesity**, London, v. 33, n. 1, p. 60-65, Apr. 2009.
DIAS, L. F. T. et al. Anti-ulcer and antioxidant activity *of Baccharis Trimera* (Less) DC (Asteraceae). **Revista Brasileira de Farmacologia**, São Paulo, v. 19, n. 1, p. 309-314, jan./mar. 2009.
DJERIDANE, A. et al. Antioxidant activity of some Algerian medicinal plant extracts containing phenolic compounds. **Food Chemistry**, London, v. 97, n. 4, p. 654-660, Aug. 2006.
FREEDMAN, D. S. et al. Cardiovascular risk factors and excess adiposity among overweight children and adolescents: the Bogalusa heart study. **Journal of Pediatrics**, New York, v. 150, n. 1, p. 12-17, jan. 2007.
GENOVESE, L. I.; LAJOLO, F. M. Trypsin inhibitory activity of beans (*Phaseolus vulgaris* L.): critical evaluation of determination methods.
Arquivos Latinos Americano de Nutricao, Caracas, v. 51, n. 4, p. 386-394, 2001.
GIGANTE, D. P. et al. Prevalence of overweight and obesity and associated factors. **Revista de Saúde Pública**, São Paulo, v. 43, n. 2, p. 83-89, nov. 2009.
GUH, D. P. et al. The incidence of co-morbidities related to obesity and overweight: a systematic review and meta-analysis. **BMC Public Health**, London, v. 9, n. 88, Mar. 2009. Available at: <http://www.ncbi.nlm.nih.gov/pubmed/19320986>. Accessed on: Nov. 12, 2012.
HALIWEELL, B. Free radicals and antioxidants: quo vadis? **Trends in Pharmacological Sciences,** London, v. 32, n. 3, p. 125-130, Mar. 2011.
HALLIWELL, B. **Free radicals and other reactive species in disease**. In: ENCYCLOPEDIA de ciências da vida. Coimbra: Natureza, 2001. Available at: <http://www.els.net/WileyCDA/ElsArticle/refId-a0002269.html>. Accessed on: Nov. 12, 2012.
HALLIWELL, B.; GUTTERIDGE, J. M. C.; CROSS, C. E. Free radicals, antioxidants, and human disease: where are we now? **Journal of Laboratory and Clinical Medicine,** Saint Louis, v. 119, n. 6, p. 598-620, June 1992.
HAMAMIZU, K. et al. Effects of *Garcinea cambogia* extract on serum leptin and insulin in rats. **Fitoterapia**, Amsterdam, v. 74, n. 3, p. 267-273, abr. 2003a.
HAMAMIZU, K. et al. Effects of *Garcinea cambogia* extract on serum sex hormones in overweight subjects. **Fitoterapia**, Amsterdam, v. 79, n. 4, p. 255261, June 2008.
HAMAMIZU, K. et al. Effects of *Garcinea cambogia* (Hidroxycitrici Acid) on viceral fat accumulation: a double-blind, randomized, placebo-controlled trial. **Current Therapeutic Research**, New York, v. 64, n. 8, p. 551-567, Sept./Oct. 2003b.
HE, Q. et al. Quality assurance and safety in the processing of aloe vera gel juice. **Food Control**, Guildford, v. 16, n. 2, p. 95-104, Feb. 2005.
HENRY, J. A.; PANDIT, A. Perspective on biomaterials used in the surgical treatment of morbid obesity. **Obesity Reviews**, Oxford, v. 10, n. 3, p. 324-332, May 2009.
BRAZILIAN INSTITUTE OF GEOGRAPHY AND STATISTICS. Prisoner Index Coordination. **Pesquisa de ornamentos familiares 2008-2009:** análise da disponibilidade domiciliar e estado nutricional no Brasil. Rio de Janeiro, 2009. 80 p.
KANDRA, L. et al. Enzymatic synthesis of a novel a-amylase inhibitor: acarviosinyl-

isomatosyl-spiro-thio-hydantoin. **Carbohydrate Research**, Amsterdam, v. 340, n. 7, p. 1311-1317, May 2005.
KAYODE, O. A. et al. Effects of crude ethanolic extract of *Garcinea cambogia* on the reproductive system of male wistar rats (*Rattus novergicus*). **African Journal of Biotechnology**, Grahamstown, v. 6, n. 10, p. 1236-1238, May 2007.
KIM, K. Y. et al. *Garcinea cambogia extract* ameliorates visceral adiposity in C57BL/6J mice fed on a high-fat diet. **Bioscience Biotechnology Biochemistry**, Tokyo, v. 72, n. 7, p. 1772-1780, July 2008.
KOVACS, E. M. R.; WESTERTERP-PLATENGA, M. S. Effects of hydroxycitric acid on net fat synthesis as de novo lipogenesis. **Physiology & Behavior**, Philadelphia, v. 88, n. 4, p. 371-381, July 2006.
KWON, Y. I.; APOSTOLIDIS, E.; SHETTY, K. Inhibitory potential of wine and tea against a-amylase and a-glucosidase for the control of hyperglycemia associated with type 2 diabetes. **Journal of Food Biochemistry**, Westport, v. 32, n. 1, p. 15-31, Feb. 2006.
LORENZI, H.; MATOS, A. F. J. **Plantas medicinais no Brasil:** nativas e exóticas. São Paulo: Instituto Plantarum de Estudos da Flora, 2002. 441 p.
MAHAN, L. K.; ESCOTT-STUMP, S. **Krause's food, nutrition, and diet therapy**. 12th ed. Philadelphia: WB Saunders, 2008. 1321 p.
MARCELLO, C. M. Evaluation of the anti-ulcer activity of *Simaba Ferruginea* ST. HIL. (Simaroubaceae). In: SIMPOSIO DE PLANTAS MEDICINAIS DO BRASIL, 17., 2002, Cuiaba. **Proceedings...** Cuiabá: UFMT, 2002. 1 CD-ROM.
MAYER, M. A. et al. Recent advances in obesity pharmacotherapy. **Current Clinical Pharmacology**, Sharjah, v. 4, n. 1, p. 53-61, Jan. 2009.
MCDOUGALL, G. J. et al. Red wine anthocyanins: their stability under simulated gastrointestinal digestion. **Phytochemistry**, Saint Paul, v. 66, n. 21, p. 2540-2548, Nov. 2005.
MENDES, F. R.; TABACH, R.; CARLINI, E. A. Evaluation of *Baccharis trimera* and *Davilla rugosa* in adaptogenic activity tests. **Phytotherapy Research**, London, v. 21, n. 6, p. 517-522, June 2007.
MOHAMED, E. A. K. Antidiabetic, antihypercholesterolemic and antioxidative effect of *Aloe Vera* gel extract in alloxan-induced diabetic rats. **Australian Journal of Basic and Applied Sciences**, Amman, v. 5, n. 11, p. 1321-1327, 2011.
MORAES, L. D.; SOUSA, O. V. Qualitative and quantitative assessments of secondary metabolite variation in *Tournefortia paniculata* Cham (Boraginaceae). **Revista Brasileira de Biociências**, Porto Alegre, v. 5, n. 2, p. 1032-1034, July 2007.
MORO, C. O.; BASILE, G. Obesity and medicinal plants. **Fitoterapia**, Amsterdam, v. 71, n. 1, p. 73-82, aug. 2000.
MUHAMMAD, I. et al. A new antimalarial quassinoid from *Simabayama orinocensis*. **Journal of Natural Products**, Washington, v. 67, n. 5, p. 772-777, May 2004.
MUKHERJEE, M. Human digestive and metabolic lipases: a brief review. **Journal of Molecular Catalysis B: Enzymaticy,** Oxford, v. 22, n. 5, p. 369376, July 2003.
NAKASUGI, T.; KOMAI, K. Antimutagens in the popular Brazilian medicinal plant carqueja (*Baccharis trimera* Less.). **Journal of Agricultural and Food Chemistry**, Washington, v. 46, n. 7, p. 2560-2564, June 1998.
NONINO-BORGES, C. B. et al. Clinical treatment of obesity. **Medicina**, Ribeirão Preto, v. 39, n. 2, p. 246-253, apr./jun. 2006.
NOOR, A. et al. Antidiabetic activity of *Aloe vera* and histology of organs in streptozotocin-induced diabetic rats. **Current Science**, Bangalore, v. 94, n. 8, p. 1078, Apr. 2008.
OBIRO, W. C.; ZHANG, T.; JIANG, B. The nutraceutical role of a-amylase inhibitor from *Phaseolus vulgaris.* **British Journal of Nutrition,** Cambridge, v. 100, n. 1, p. 1-12, July 2008.
OLIVEIRA, A. C. P. et al. Effect of extracts and fractions of *Baccharis trimera* and *Syzygium cumini* on glycemia in diabetic and non-diabetic rats. **Journal of Ethnopharmacology**, Lausanne, v. 102, n. 3, p. 465-469, dec. 2005.
OLIVEIRA, S. Q. et al. Antioxidant activity of *Baccharis articulata* extracts: isolation of a new compound with antioxidant activity. **Free Radical Research**, Oxford, v. 37, n. 5, p. 555-559, May 2003.
PAN-AMERICAN HEALTH ORGANIZATION. **Chronic degenerative diseases and obesity:** global strategy on healthy eating, physical activity and health. Brasília: PAHO/WHO, 2003. 60

p.
OTA, M. et al. Action of a-D-glucosidase from *Aspergillus niger* towards dextrin and starch. **Carbohydrate Polymers**, Oxford, v. 78, n. 2, p. 287-291, Apr. 2009.
PADUA, B. C. et al. Antioxidant properties of *Baccharis trimera* in neutrophils of fisher rats. **Journal of Ethnopharmacology**, Lausanne, v. 129, n. 3, p. 381-386, June 2010.
PARK, M. Y.; LEE, K. S.; SUNG, M. K. Effects of dietary mulberry, Korean red ginseng and banaba on glucose homeostasis in relation to mRNA expressions of PPAR-a, PPAR-y and LPL. **Life Sciences**, Oxford, v. 77, n. 26, p. 3344-3354, Nov. 2005.
PERON, A. P. et al. Mutagenic evaluation of the medicinal plants *Baccharis trimera* Less. and *Solanum melongena* L. in bone marrow cells of Wistar rats. **Revista Brasileira de Biociências**, Porto Alegre, v. 6, n. 2, p. 127-130, Apr./Jun. 2008.
PERUSSE, L.; BOUCHARD, C. Genotype-environment interaction in human obesity. **Nutrition Reviews**, New York, v. 57, n. 5, p. S31-S37, May 1999.
PINHO, D. S. et al. Evaluation of the mutagenic activity of *Baccharis trimera* (Less.) DC. infusion in *Allium cepa* test and chromosomal aberration test in human lymphocytes. **Revista Brasileira de Farmacognosia**, São Paulo, v. 20, n. 2, p. 165-170, Apr./May 2010.
POPKIN, B. M.; GORDON-LARSEN, P. The nutrition transition: worldwide obesity dynamics and their determinants. **International Journal of Obesity**, London, v. 28, n. 3, p. 2-9, nov. 2004.
RAJASEKARAN, S. K. et al. Beneficial effects of *Aloe vera* leaf gel extract on lipid profile in rats with streptozotocin diabetes. **Clinical and Experimental Pharmacology and Physiology**, Carlton, v. 33, n. 3, p. 232-237, Mar. 2006.
RAJASEKARAN, S. K. et al. Hypoglycemic effect of *Aloe vera* gel on streptozotocin induced diabetes in experimental rats. **Journal of Medicinal Food**, New Rochelle, v. 7, n. 1, p. 61-66, 2004.
RAJASEKARAN, S. K.; SIVAGNAMAN, S.; SUBRAMANIAN, S.
Modulatory effects of *Aloe vera* leaf gel extract on oxidative stress in streptozotocin-treated rats. **Journal of Pharmacy and Pharmacology**, London, v. 57, n. 2, p. 241-246, Feb. 2005.
RAMALHO, V. C.; JORGE, N. Antioxidants used in oils, fats and fatty foods. **Química Nova**, São Paulo, v. 29, n. 4, p. 755-760, jul./ago. 2006.
RAYALAM, S.; DELLA-FERA, M. A.; BAILE, C. A. Phytochemicals and regulation of the adipocyte life cycle. **Journal Nutritional Biochemistry**, Philadelphia, v. 19, n. 11, p. 717-726, Nov. 2008.
RODRIGUES, C. R. F. et al. Genotoxic and antigenotoxic properties of *Baccharis trimera* in rats. **Journal of Ethnopharmacology**, Lausanne, v. 125, n. 1, p. 97-101, aug. 2009.
RODRIGUES, H. G. et al. Nutritional supplementation with natural antioxidants: effect of rutin on HDL-cholesterol concentration. **Revista de Nutri^ao**, Campinas, v. 16, n. 3, p. 315-320, jul./set. 2003.
ROVER JÚNIOR, L.; HOEHR, N. F.; VELLASCO, A. P. Antioxidant system involving the glutathione metabolic cycle associated with electroanalytical methods in the evaluation of stress. **Quimíca Nova,** São Paulo, v. 24, n. 1, p. 112-119, jan./feb. 2001.
SAITO, M. et al. High dose of *Garcínea cambogia* is effective in suppressing fat accumulation in developing obese male Zucker rats but highly toxic to the testis**. Food and Chemical Toxicology**, Oxford, v. 43, n. 3, p. 411-419, Mar. 2005.
SANTOS, A. C. S. et al. *Garcinia cambogia*: a plant species as a therapeutic resource against obesity? **Natureza Online**, Santa Teresa, v. 5, n. 1, p. 3743, 2007.
SCHACHTER, M. Kallikreins (kininogenases): a group of serine proteases with bioregulatory actions. **Pharmacological Reviews**, Bethesda, v. 31, n. 1, p. 1-17, 1980.
SGARBIERI, V. C. **Alimentado e nutrido:** fator saúde e desenvolvimento. São Paulo: Almed, 1987. 387 p.
SHARMA, N.; SHARMA, V. K.; SEO, S. Y. Screening of some medicinal plants for anti-lipase activity. **Journal of Ethnopharmacology**, Lausanne, v. 97, n. 3, p. 453-456, Mar. 2005.
SHI, Y.; BURN, P. Lipid metabolic enzymes: emerging drug targets for the treatment of obesity. **Nature Reviews Drug Discovered**, Washington, v. 3, n. 8, p. 695-710, Aug. 2004.
SILVA, E. M. **Inhibitory action of cerrado plant extracts on alpha-amylases with emphasis on Kielmeyera coriacea**. 2008. 141 p. Dissertation (Master's in Health Science) - University of Brasília, Brasília, 2008.

SILVA, M. P.; SILVA, M. P. A. P. Antinutritional factors: protease inhibitors and lectins. **Revista de Nutrido**, Campinas, v. 13, n. 1, p. 3-9, jan./abr. 2000.
SIMAO, A. A.; CORREA, A. D.; CHAGAS, P. M. B. Inhibition of digestive enzymes by aqueous extracts of medicinal plants used in the treatment of obesity. **Journal of Medicinal Plants Research**, Lagos, v. 6, n. 47, p. 58265830, dec. 2012.
SOARES, S. E. Phenolic acids as antioxidants. **Revista de Nutrido**, Campinas, v. 15, n. 1, p. 71-81, jan. 2002.
BRAZILIAN SOCIETY OF ENDOCRINOLOGY AND METABOLISM. **Obesity**. Rio de Janeiro, 2010. Available at: <http://www.endocrino.org.br/busca/obesidade>. Accessed on: September 20, 2010.
SORENSEN, J. F. et al. Potential role of glycosidase inhibitors in industrial biotechnological applications. **Biochimica et Biophysica Ata**, Alberta, v. 1696, n. 2, p. 275-287, Feb. 2004.
SOUZA, S. P. et al. Inhibition of pancreatic lipase by extracts of *Baccharis trimera* (Less.) DC., Asteraceae: evaluation of antinutrients and effect on glycosidases. **Revista Brasileira de Farmacognosia**, Curitiba, v. 21, n. 3, p. 450-455, May/Jun. 2011.
STANlC, S. Antigenotoxic effect of *Aloe vera* gel on the mutagenic action of ethyl methanesulfonate. **Archives of Biological Sciences**, Belgrade, v. 59, n. 3, p. 223-226, 2007.
STURBELLE, R. T. et al. Evaluation of the mutagenic and antimutagenic activity of *Aloe vera* in the *Allium cepa* test and micronucleus test in binucleated human lymphocytes. **Revista Brasileira de Farmacognosia**, Curitiba, v. 20, n. 3, p. 409-415, jun./jul. 2010.
SUNDQUIST, J.; JOHANSSON, S. E. The influence of socioeconomic status, ethnicity and lifestyle on body mass index in a longitudinal study. **International Journal of Epidemiology**, Oxford, v. 27, n. 1, p. 57-63, feb. 1998.
TALMADGE, J. et al. Fractionation of Aloe vera L. inner gel, purification and molecular profiling of activity. **International Immunopharmacology**, Amsterdam, v. 4, n. 14, p. 1757-1773, Dec. 2004.
TEIXEIRA, S. S. **Herbal medicines and industrialized plant drugs and medicinal plants officialized by the Ministry of Health in Brazil:** health regulations, scope and quality of pre-clinical and clinical studies. 2013. 337 p. Thesis (Doctorate in Public Health) - Sergio Arouca National School of Public Health, Oswaldo Cruz Foundation, Rio de Janeiro, 2013.
TOMA, W. et al. Antiulcerogenic activity of four extracts obtained from the bark wood of *Quasia amara* L. (Simaroubacea). **Biological and Pharmaceutical Bulletin**, Tokyo, v. 25, n. 9, p. 1151-1155, sep. 2002.
TUCCI, S. A.; BOYLAND, E. J.; HALFORD, J. C. G. The role of lipid and carbohydrate digestive enzyme inhibitors in the management of obesity: a review of current and emerging therapeutic agents. **Diabetes, Metabolic Syndrome and Obesity: Targets and Therapy**, Manchester, v. 3, p. 125-143, May 2010.
UDANI, J. K. et al. Reduction of the glycemic index of white bread using a white bean extract. **Nutrition**, New York, v. 8, n. 52, p. 1-5, Oct. 2009.
VALVERDE, J. M. et al. New edible coating based on *Aloe vera* gel to maintain the quality and safety of table grapes. **Journal of Agricultural and Food Chemistry**, Washington, v. 53, n. 20, p. 7807-7813, Oct. 2005.
VERDI, L. G. et al. The *genus Baccharis* (Asteraceae): chemical, economic and biological aspects. **Química Nova**, São Paulo, v. 28, n. 1, p. 85-94, jan./feb. 2005.
WANDERLEY, E. N.; FERREIRA, V. A. Obesity: a plural perspective. **Ciência & Saúde Coletiva**, Rio de Janeiro, v. 15, n. 1, p. 185-194, jan. 2010.
WANG, J. R. et al. Molecular evolution of dimeric a-amylase inhibitor genes in wild wheat and their ecological association. **BMC Evolutionary Biology**, London, v. 8, n. 91, p. 91-105, Mar. 2008.
WANG, Y. et al. Trends in obesity and underweight in older children and adolescents in the United States, Brazil, China, and Russia. **American Journal of Clinical Nutrition**, Bethesda, v. 75, n. 6, p. 971-977, June 2002.
WIKIPEDIA. **Moderdiet**. Available at: <http://www.wikipédia/moderdiet.htm>. Accessed on: September 28, 2010.
WORLD HEALTH ORGANIZATION. **WHO global database on body mass index**. Geneva, 2010. Available at:
<http://apps.who.int/bmi/index.jsp>. Accessed on: June 10, 2012.

YILDIRIM, A.; MAVI, A.; KARA, A. A. Determination of the antioxidant and antimicrobial activities of extracts of Rumex crispus L. **Journal of Agricultural and Food Chemistry,** Washington, v. 49, n. 8, p. 4083-4089, Aug. 2001.

YUN, H. U.; JUAN, X. U.; QIUHUI, H. U. Evaluation of the antioxidant potential of *Aloe vera* (*Aloe barbadensisMiller*) extracts. **Journal of Agricultural and Food Chemistry,** Washington, v. 51, n. 26, p. 7788-7791, dec. 2003.

YUNES, R. A.; CALIXTO, J. B. **Plantas medicinais sob a ótica da química medicinal moderna**. Chapecó: Argos, 2001. 120 p.

ASSOCIATION OF OFFICIAL ANALYTICAL CHEMISTS. **Official methods of analysis of the association of the analytical chemists**. 17. ed. Washington, 2005.

BACCOU, J. C.; LAMBERT, F.; SAUVAIRE, Y Spectrophotometric method for the determination of total steroid sapogenin. **Analyst**, London, v. 102, n. 1215, p. 458-465, 1977.

BALASUNDRAM, N.; SUNDRAM, K.; SAMMAR. S. Phenolic compounds in plants and agro-industrial by-products: Antioxidant activity, occurrence and potential uses. **Food Chemistry**, v. 68, p. 191-203, 2006.

CATALDO, D. A.; MAROON, M.; SCHRADER, L. E.; YOUNGS, V. L. Rapid colorimetric determination of nitrate in plant tissutive by nitration of salicylic acid. **Communications in Soil Science and Plant Analysis**, New York, v. 6, n. 1, p. 71-80, 1975.

ERLANGER, B. F.; KUKOWSKY, N.; COHEN, W. The preparation and properties of two new chromogenic substrates of trypsin. **Archives of Biochemistry and Biophysics**, New York, v. 95, p. 271-278, Nov. 1961.

FAQUIN, V.; ANDRADE, A. T. **Nitrate accumulation in vegetables and human health.** Lavras: UFLA/FAEPE, 2004, 88 p.

FRANCIS, G.; KEREM, Z.; MAKKAR, H. P.; BECKER, K. The biological action of saponins in animal systems: a review. **British Journal of Nutrition**, v. 88, n. 6, p. 587-605, 2002.

FREITAS, M. S. M.; MARTINS, M.A.; CARVALHO, A. J. C.; CARNEIRO, R. F. V. Growth and production of total phenols in carqueja [*Baccharis trimera* (Less.) DC.] in response to inoculation with arbuscular mycorrhizal fungi, in the presence and absence of mineral fertilization. **Revista Brasileira de Plantas Medicinais**, v. 6, n. 3, p. 30-34, 2004.

GUH, D. P.; ZHANG, W.;BANSBACK, N.; AMARSI, Z.; BIRMINGHAM, C. L.; ANIS, A. H. The incidence of obesity- and overweight-related comorbidities: A systematic review and meta-analysis. **Bmc Public Health,** v. 9, n. 88, p. 1-20, 2009.

JANTAN, I.; FARRA, A. J.; FADLINA, C. S.; KHALID, R. Inhibitory effects of *Garcinia* species extracts on human low-density lipoprotein peroxidation and platelet aggregation in relation to their total phenolic contents. **Journal of Medicinal Plants Research**, v. 5, n. 13, p. 2699-2709, 2011.

KAKADE, M. L.; SIMONS, N.; LIENER, I. E. Determination of trypsin inhibitory activity of soybean products: A collaborative analysis of an improved procedure. **Cereal Chemistry**, Saint Paul, v. 51, p. 376-382, 1974.

HIDALGO, M. E.; FERNÁNDEZ, E.; QUILHOT, W.; LISSI, E. Antioxidant activity of depsides and depsidones. **Phytochemistry**, v. 37, n. 6, p. 1585-1587, 1994.

HEANEY, R. P. Normalizing calcium intake: projected population effects for body weight. **Journal of Nutrition**, v. 133, n. 1, p. 268-270, 2003.

LIU, S.; WILLETT, W. C.; MANSON, J. E.; HU, F. B.; ROSNER, B.; COLDITZ, G. Relationship between changes in dietary fiber and grain product intake and changes in weight and development of obesity among middle-aged women. **American Journal Clinical Nutrition**, v. 78, n. 5, p. 920-927, 2003.

LOURES, A.; JOKL, L. Microtechnique for the determination of oxalic acid in leaves and derivatives. In: **ENCONTRO NACIONAL DE ANALISTAS DE ALIMENTOS**, 6., 1990, Curitiba. Abstracts... Curitiba: Paraná Institute of Technology, 1990. p 59.

MAHAN, L. K.; ESCOTT-STUMP, S. **Krause's food, nutrition, and diet therapy**. 12th ed. Philadelphia: WB Saunders, 2008.

MALAVOLTA, E.; VITTI, G. C.; OLIVEIRA, S. A. Evaluation of plant nutritional status. Piracicaba: **Potafos,** 1997. 319 p.

MATOS, F. J. A. **Introdugão a phitoquímica experimental**. 2.ed. Fortaleza: UFC, 1997. 141 p.

MAYER, M. A.; HOCHT, C.; PUYO, A.; TAIARA, C. A. Recent advances in obesity pharmacotherapy. **Current Clinical Pharmacology**, v. 4, n. 1, p. 53-61, 2009.

MELLO, V, D.; LAAKSONEN, D, E. Dietary fibers: current trends and health benefits in metabolic syndrome and type 2 diabetes. **Arquivos Brasileiros de Endocrinologia e Metabologia**, v. 53, n. 5, p. 509-518 2009.
MONIRUZZAMAN, M.; BEGUM, R.; SOHEL, A.; AMRITA, B.; IBRAHIM, K.; SIEW, H. *In vitro* antioxidant effects of *Aloe barbadensis* Miller extracts and the potential role of these extracts as antidiabetic and antilipidemic agents in streptozotocin-induced type 2 diabetic model rats. **Molecules**, v. 17, n. 11, p. 12851-12867, 2012.
MONTEIRO, J. M.; ALBUQUERQUE, U. P.; ARAÚJO, E. L. Taninos: Uma Abordagem da química a ecologia. **Química Nova**, v. 28, n. 5, p. 892-896, 2005.
MOORE, L. L.; SINGER, M. R.; BRADLEE, M. L.; ELLISON, R. C. Dietary predictors of excess body fat acquisition during childhood. **Circulation**, v. 197, n. 7, p. 5, 2004.
MORAES, L. D.; SOUZA, O. V. Avaliacões Qualitativas e Quantitativas da Varia^ao **de** Metabólitos Secundários em *Tournefortia paniculata* Cham (Boraginaceae), **Revista Brasileira de Biociencias,** Porto Alegre, v. 5, n. 2, p. 1032-1034, 2007.
NEAMATI, N.; HONG, H.; MAZUMDER, A.; WANG, S.; SUNDER, S.; NICKLAUS, M. C.; MILNE, G. W. A.; PROKSA, B.; POMMIER, Y. Depsides and Depsidones as HIV-1 Integrase Inhibitors: Discovery of new inhibitors through 3D database searching. **Journal of Medicinal Chemistry,** v. 40, n. 6, p. 942-951, 1997.
OLIVEIRA, C. B.; COMUNELLO, L. N.; LUNARDELLI, A.; AMARAL, R. H.; PIRES, M. G. S.; SILVA, G. L.; MANFREDINI, V.; VARGAS, C. R.; GNOATTO, S. C. B.; OLIVEIRA, J. R.; GOSMANN, G. Enriched phenolic extract of *Baccharis trimera* presents anti-inflammatory and antioxidant activities. **Moléculas**, v. 17, n. 1, p. 11131123, 2012.
OKUYAMA, E.; UMEYAMA, K.; YAMAZAKI, M.; KINOSHITA Y.; YAMAMOTO, Y. Usnic acid and difractic acid as analgesic and antipyretic components of *Usnea diffracta.* **Planta Med,** v. 61, p. 113115, 1995.
PEREIRA, R. J.; CARDOSO. M. G. Plant secondary metabolites and antioxidant benefits. **Revista de Biotecnologia e Biodiversidade**, v. 3, n. 4, p. 146-152, 2012.
PIETTA, P. G. Flavonoids as antioxidants. **Journal of Natural Products**, v. 63, n. 7, p. 1035-1042, 2000.
R core development team. R: **A language and environment for statistical computing**. Vienna: R Foundation for Statistical Computing 2011.
RIQUE, A. B. R.; SOARES, E. A.; MEIRELLES, C. M. Nutrido e exercício na prevenção e controle das doenças cardiovasculares. **Revista Brasileira de Medicina do Esporte**, v. 8, n. 6, p. 1-11, 2002.
RODRIGUES, C. R. F.; DIAS, J. H.; DE MELO, R. N.; RITCHER, R. F.; PICADA, J. N.; FERRAZ, A. B. Genotoxic and antigenotoxic properties of *Baccharis trimera* in mice. **Journal of Ethnopharmacology**, v. 125, n. 1, p. 97-101, 2009.
SCALBERT, A.; JOHNSON, I. T.; SALTMARSH, M. Polyphenols: antioxidants and beyond. **American Journal of Clinical Nutrition**, v. 81, n. 1, p. 215-217, 2005.
SIENER, R.; HUNOW, R.; SEIDLER, A.; VOSS, S.; HESSE, A. Oxalate content in species of the families Polygonaceae, Amaranthaceae and Chenopodiaceae. **Food Chemistry**, v. 98, n. 2, p. 220-224, 2005.
SIMAO, A. A.; CORREA, A. D.; CHAGAS, P. M. B. Inhibition of digestive enzymes by aqueous extracts of medicinal plants used in the treatment of obesity. **Journal of Medicinal Plants Research**, v. 6, n. 47, p. 5826-5830, 2012.
SIMAO, A. A.; LAGE, F. F.; CHAGAS, P. M. B.; FRAGUAS, R. M.; FREIRE, J. M.; MARQUES, T. R.; CORREA, A. D. Antioxidants of Medicinal Plants Used in the Treatment of Obesity. **European Journal of Medicinal Plants**, v. 3, n. 3, p. 429-443, 2013.
SIMOES, C. M.; SHENKEL, E. P.; GOSMAN, G.; MELLO, J. C. P.;
MENTZ, L. A.; PETROVICK, P. R. **Farmacognosia: da planta ao medicamento**. Porto Alegre: Ed. Da UFRGS; Florianópolis: Ed.5 Da UFSC, 2007. 1102 p.
SOUZA, S. P.; PEREIRA, L. L. S.; SOUZA, A. A.; SANTOS, C. D. Inhibition of pancreatic lipase by extracts of *Baccharis trimera* (Less.) DC. Asteraceae: evaluation of antinutrients and effect on glycosidases. **Revista Brasileira de Farmacognosia**, v. 21, n. 3, p. 450-455, 2011.
SUBHASHINI, N.; NAGARAJAN, G.; KAVIMANI, S. *In vitro* antioxidant and anticholinesterase activities of *Garcinia combogia.* **International Journal of Pharmacy and Pharmaceutical Sciences**, v. 13, n. 3, p. 129-132, 2011.

ST-ONGE, M. P. Dietary fats, teas, dairy products and nuts: potential functional foods for weight control? **American Journal of Clinical Nutrition**, v. 81, n. 1, p. 7-15, 2005.
VASQUEZ, B.; AVILA, B.; SEGURA, D.; ESCALANTE B. Anti-inflammatory activity of *Aloe vera* gel extracts. **Journal of Ethnopharmacology**, v. 55, n. 1, p. 69-75, 1996.
WORLD HEALTH ORGANIZATION. **WHO series no. 50 on food additives**. Safety evaluation of certain food additives. Fifty-ninth report of the Joint FAO/WHO Committee on Food Additives. Geneva. 2003.
WORLD HEALTH ORGANIZATION. **WHO global database on body mass index**. WHO: Geneva, 2010.
ZEMEL, M. B.; THOMPSON, W.; MILSTEAD, A.; MORRIS, K.; CAMPBELL, P. Calcium and dairy acceleration of weight and fat loss during energy in obese adults. **Obesity Research**, v. 12, n. 4, p. 582-590, 2004.
AOAC (2005). Official methods of analysis of the association of the analytical chemists (17^{ed}) Association of Official Analytical Chemists. Washington, DC. USA.
Bhutani KK, Birari R, Kapat K (2007). Potential anti-obesity and lipid-lowering natural products: a review. Nat. Prod. Commun. 2: 331-348.
Bray GA (2004). Medical consequences of obesity. J. Clin. Endocrinol. Metab. 89(6): 2583-2589.
Celleno L, Tolaine MV, Damore A, Perricone MV, Preuss HG (2007). A dietary supplement containing standardized Phaseolus vulgaris extract influences body composition in overweight men and women. Int. J. Med. Sci. 4(1): 45-52.
Chen X, Xu G, Li X, Li Z, Ying H (2008). Purification of an a-amylase inhibitor in a polyethylene glycol/fructose-1,6-biphosphate trisodium salt aqueous two-phase system. Process Biochesm. 43(7): 765-768.
Erlanger BF, Kukowsky N, Cohen, W (1961). Preparation and properties of two new chromogenic trypsin substrates. Arch. Biochem. Biophys. 95(2): 271-278.
Friedman M, Brandon DL (2001). Nutritional and health benefits of soy proteins. J. Agric. Food Chem. 49(3): 1069-1086.
Kandra L (2003). a-amylases of medicinal and industrial importance. J. Mol. Struct. 487: 666-667.
Kwon YI, Apostolidis E, Shetty K (2006). Inhibitory potential of wine and tea against a-amylase and a-glucosidase for management of hyperglycemia linked to type 2 diabetes. J. Food Biochem. 32(1): 15-31.
McDougall GJ, Fiffe S, Dobson P, Stewart D (2005). Red wine anthocyanins - Their stability under simulated gastrointestinal digestion. Phytochemistry. 66(21): 2540-2548.
Noelting G, Bernfeld P (1948). On amylolytic enzymes - III. La ß-amylase: dosage d'activité et contrôle de l'absence d'a- amylase. Helv. Chim. Ata. 31(1): 286-290.
Obiro WC, Zhang T, Jiang B (2008). The nutraceutical role of a-amylase inhibitor from *Phaseolus vulgaris*. Br. J. Nutr. 100(1): 1-12.
Ota M, Okamoto T, Hoshino W, Wakabayashi H (2009). Action of a-D-glucosidase from Aspergillus niger on dextrin and starch. Carbohydrate Polymers. 78:287-291.
Pereira LLS, Pereira CA, Santos CD, Marques TR, Silva MC (2010). Standardization of a-amylase inhibitor protein extraction from white beans. Nat. Sci. Nat. 32(2): 51-59.
Pereira CA, Pereira LLS, Corrêa AD, Chagas PMB, Souza SP, Santos CD (2011). Inhibition of digestive enzymes by commercial powdered extracts of *Hoodia gordonii*. Rev. Bras. Biociên. 9(3): 265269.
Sharma N, Sharma VK, Seo SY (2005). Screening of some medicinal plants for anti-lipase activity. J. Ethnopharmacol. 97(3): 453-456.
Souza SP, Pereira LLS, Souza AA, Santos CD (2011). Inhibition of pancreatic lipase by extracts of *Baccharis trimera* (Less.) DC. Asteraceae: evaluation of antinutrients and effect on glycosidases. Rev. Bras. Farmacogn. 21(3): 450-455.
Sugimoto S, Nakamura S, Yamamoto S, Yamashita C, Oda Y, Matsuda H, Yoshikawa M (2009). Brazilian natural medicines. III. Structures of triterpene oligoglycosides and lipase inhibitors from Mate, *Ilex paraguarienses* leaves. Biol. Pharm. Bull. 57(3): 257261.
THE UNITED STATES PHARMACOPOEIA (2005). The national formulary NF 18 (Pharmacopeial Convention Ing). Rockvile.
Tucci SA, Boyland EJ, Halford JCG (2010). The role of lipid and carbohydrate digestive enzyme inhibitors in the management of obesity: a review of current and emerging therapeutic agents.

Diabetes, Metabolic Syndrome and Obesity: Targets and Therapy. 3: 125-143.
Udani JK, Singh BB, Barret ML, Preuss HG (2009). Reducing the glycemic index of white bread using a white bean extract. Nutrition. 8: 1-5.
Viegas-Jr C, Bolzani VS, Barreiro EJ (2006). Natural products and modern medicinal chemistry. Chem. Nova. 29(2): 326-337.
Halliwell B. The paradox of antioxidants. Lancet. 2000; 355(9210):1179-1180.
Morais SM. Antioxidant action of teas and condiments most consumed in Brazil. Rev Bras Farmacogn. 2009; 19(1): 315-320.
Halliwell B. Free radicals, antioxidants, and human disease: curiosity, cause, or consequence. Lancet. 1994; 344(8924): 721-724.
Djeridane A, Yousfi M, Nadjemi B, Boutassouna D, Stocker P, Vidal N. Antioxidant activity of some Algerian medicinal plants extracts containing phenolic compounds. Food Chem. 2006; 97(4): 654-660.
Souza SP, Pereira LLS, Souza AA, Santos CD. Inhibition of pancreatic lipase by extracts of *Baccharis trimera* (Less.) DC. Asteraceae: evaluation of antinutrients and effect on glycosidases. Rev Bras Farmacogn. 2011; 21(3): 450-455.
Simão AA, Correa AD, Chagas PMB. Inhibition of digestive enzymes by aqueous extracts of medicinal plants used to treat obesity. J Med Plant Res. 2012; 6(47): 5826-5830.
Villareal DT, Apovian CM, Kushner RF, Klein F. Obesity in older adults: technical review and position statement of the American Society for Nutrition and NAASO, The Obesity Society. Am J Clin Nutr. 2005; 82(5): 923-934.
AOAC. Official methods of analysis of the association of the analytical chemists (17^{ed}) Association of Official Analytical Chemists. Washington, DC. USA. 2005.
Zhishen J, Mengcheng T, Jianming W. The determination of flavonoid content in mulberry and their superoxide radical scavenging effects. Food Chem. 1999; 64(4): 555-559.
Strohecker R, Henning HM. Vitamin analysis: proven methods. Madrid: Paz Montalvo; 1967.
Higby WK. A simplified method for determination of some the carotenoid distribution in natural and carotene-fortified orange juice. J Food Sci. 1962; 27(1): 42-49.
Nagata M, Yamashita I. A simple method for the simultaneous determination of chlorophyll and carotenoids in tomato fruit. J Japan Soc Food Sci Technol. 1992; 39(10): 925-928.
Rufino MSM, Alves RS, Brito ES, Filho JM, Moreira AVB. Determination of total antioxidant activity in fruit using the ß-carotene/linoleic acid method. Fortaleza: Embrapa Agroindústria tropical. 2006.
Re R, Pellegrini N, Proteggente A, Pannala A, Yang M, Rice-Evans C. Antioxidant activity applying an improved ABTS cation radical scavenging assay. Free Radic Biol Med. 1999; 26(9/10): 1231-1237.
Rufino MSM, Alves RS, Brito ES, Morais SM. Determination of total antioxidant activity in fruit using ABTS free radical scavenging^{+} . Fortaleza: Embrapa Agroindústria tropical. 2007.
R core team. R: A language and environment for statistical computing. Vienna: R Foundation for Statistical Computing 2012. ISBN 3-900051-07-0. Available: http://www.R-project.org/.
Freitas MSM, Martins MA, Carvalho AJC, Carneiro RFV. Growth and production of total phenols in carqueja [*Baccharis trimera* (Less.) DC.] in response to inoculation with arbuscular mycorrhizal fungi, in the presence and absence of mineral fertilization. Rev Bra Plant Med. 2004; 6(3): 30-34.
Oliveira CB, Comunello LN, Lunardelli A, Amaral RH, Pires MGS, Silva GL, Manfredini V, Vargas CR, Gnoatto SCB, Oliveira JR, Gosmann G. Enriched Phenolic Extract of *Baccharis trimera* Presents Anti-inflammatory and Antioxidant Activities. Molecules. 2012; 17(1), 1113-1123.
Moniruzzaman M, Begum R, Sohel A, Amrita B, Ibrahim K, Siew H. *In vitro* antioxidant effects of *Aloe barbadensis* Miller extracts and the potential role of these extracts as antidiabetic and antilipidemic agents in streptozotocin-induced type 2 diabetic model rats. Molecules. 2012; 17(11): 12851-12867.
Subhashini N, Nagarajan G, Kavimani S. *In vitro* antioxidant and anticholinesterase activities of *garcinia combogia.* Int J Pharm Pharm Sci. 2011; 13(3): 129-132.
Jantan I, Farra AJ, Fadlina CS, Khalid R. Inhibitory effects of *Garcinia* species extracts on human low-density lipoprotein peroxidation and platelet aggregation in relation to their total phenolic content. J Med Plant Res. 2011; 5(13): 2699-2709.

Ghimire BK, Seong ES, Kim EH, Ghimeray AK, Yu CY, Ghimire BK, Chung MA. Comparative evaluation of antioxidant activity of some medicinal plants popularly used in Nepal. J Med Plant Res. 2011; 5(10): 1884-1891.
Wojdylo A, Osmianski J, Czermerys R. Antioxidant activity and phenolic compounds in 32 selected herbs. Food Chem. 2007; 105(3): 940-949.
Gan R, Xu XR, Song FL, Kuang L, Li HB. Antioxidant activity and total phenolic content of medicinal plants associated with the prevention and treatment of cardiovascular and cerebrovascular diseases. J Med Plant Res. 2010; 4(22): 2438-2444.
Surveswaran S, Cai Y, Corke H, Sun M. Systematic evaluation of natural phenolic antioxidants from 133 Indian medicinal plants. Food Chem. 2007; 102(3): 938-953.
Balasundram N, Sundram K, Sammar S. Phenolic compounds in plants and agro-industrial by-products: Antioxidant activity, occurrence, and potential uses. Food Chem. 2006; 99(1): 191-203.
Borella JC, Fontoura A. Evaluation of the chromatographic profile and flavonoid content in samples of *Baccharis trimera* (Less.) DC. Asteraceae (carqueja) sold in Ribeirão Preto, SP, Brazil. Rev Bras Farmacogn. 2002; 12(2): 63-67.
Sumazian Y, Syahida A, Hakiman M, Maziah M. Antioxidant activities, flavonoids, ascorbic acid and phenolic content of Malaysian vegetables. J Med Plant Res. 2010; 4(10): 881-890.
Bouayed J, Piri K, Rammal H, Dicko A, Desor F, Younos C, Soulimani S. Comparative evaluation of the antioxidant potential of some Iranian medicinal plants. Food Chem. 2007; 104(1): 364-368.
Halliwell B, Rafter J, Jenner A. Health promotion by flavonoids, tocopherols, tocotrienols and other phenols: direct or indirect effects? Antioxidants or not? Am J Clin Nutr. 2005; 81(1): 2685-2765.
Kris-Etherton PM, West SG. Soy protein with or without isoflavones: in search of a mechanism of cardioprotective action. Am J Clin Nutr. 2005; 81(1): 5-6.
Marinova D, Ribarova F, Atanassova M. Total phenolics and total flavonoids in Bulgarian fruits and vegetables. J Univ Chem Technol Metall. 2005; 40(3): 255-260.
Zhao X, Aldini G, Johnson EJ, Rasmussen H, Kraemer K, Woolf H, Musaeus N. Modification of lymphocyte DNA damage by carotenoid supplementation in postmenopausal women. Am J Clin Nutr. 2006; 83(1): 163-169.
Chao JCJ, Huang CH, Wu SJ, Yang SC. Effects of ß-carotene, vitamin C and E on antioxidant status in hyperlipidemic smokers. J Nutr Biochem. 2002; 13(7): 427-434.
Rajasekaran S, Sivagnaman K, Subramanian S. Modulatory effects of *Aloe vera* leaf gel extract on oxidative stress in rats treated with streptozotocin. J Pharm Pharmacol. 2005; 57(2): 241-246.
Saritha V, Anilakumar KR, Farhath K. Antioxidant and antibacterial activity of *Aloe vera* gel extracts. Int J Pharm Biol Arch. 2010; 1(4):376-384
Yun HU, Juan XU, Qiuhui HU. Evaluation of the Antioxidant Potential of *Aloe vera* Extracts (*Aloe barbadensisMiller*). *J Agr Food Chem. 2003;* 51(26): 77887791.
Pádua BC, Silva LD, Rossoni Junior JV, Humberto JL, Chaves MM, Silva ME, Pedrosa ML, Costa DC. Antioxidant properties of *Baccharis trimera* in neutrophils of Fisher rats. J Ethnopharmacol. 2010; 129(3): 381-386.
Mendes FR, Tabach R, Carlini EA. Evaluation of *Baccharis trimera* and *Davilla rugosa* in adaptogenic activity tests. Phytother Res. 2007; 21(6): 517522.
Dias LFT, Melo ES, Hernandes LS, Bacchi EM. Anti-ulcer and antioxidant activity *of Baccharis Trimera* (Less) DC (Asteraceae). Rev Bras Farmacogn. 2009; 19(1): 309-314.
AHN, J.; LEE, H.; KIM, S.; PARK, J.; HA, T. Anti-obesity effect of quercetin is mediated by AMPK and MAPK signaling pathways. **Biochemical Biophysical Research Communications**, v. 373, n. 4, p. 545-549, 2008.
AL-HABORI, M.; AL-AGHBARI, A.; AL-MAMARY, M.; BAKER, M. Toxicological evaluation of *Catha edulis* leaves: a long-term feeding experiment in animals. **Journal Ethnopharmacology**, v. 83, n. 3, p. 209217, 2002.
AOAC. (2005). **Official methods of analysis of the association of the analytical chemists** (17[ed]) Association of Official Analytical Chemists. Washington, DC. USA.
BALU, M., SANGEETHA, P., HARIPRIYA, D., PANNEERSELVAM, C. Rejuvenation of the antioxidant system in the central nervous system of aged rats by grape seed extract. **Neuroscience Letters**, v. 383, n. 3, p. 295-300, 2005.
BERTOLUCCI, S. K. V.; PINTO, J. B. P.; CARDOSO, M. G.; GAVILANES, M. L.;

SANTIAGO, E. J. A.; LAMEIRA, O. A. Micropropagation of *Tournefortia paniculata* Cham. 2000. **Revista Brasileira de Plantas Medicinais**, v. 3, n. 1, p. 43-49, 2000.
BIRARI, R. B.; BHUTANI, K. K. Pancreatic lipase inhibitors from natural sources: unexplored potential. **Drug Discovery Today**, v. 12, n. 19/20, p. 879-889, 2007.
COLLINS, A. R. The comet assay for DNA damage and repair: principles, applications, and limitations. **Molecular Biotechnology**, v. 26, n. 3, p. 249-261, 2004.
COSKUN, O.; KANTER, M.;, KORKMAZ, A.;, OTER, S. Quercetin, an antioxidant flavonoid, prevents and protects streptozotocin-induced oxidative stress and beta cell damage in rat pancreas. **Pharmacolical Research**, v. 51, n. 2, p. 117-123, 2005.
DUARTE-ALMEIDA, J. M., SANTOS, R. J.; GENOVESE, M. I.; LAJOLO, F. M. Evaluation of antioxidant activity using the b-carotene/linoleic acid system and the DPPH radical scavenging method. **Ciencia e Tecnología de Alimentos**, Campinas, v. 26, n. 2, p. 446-452, 2006.
GARCÍA-SAURA, M. F.; GALISTEO, M.; VILLAR, I. C.; BERMEJO, A.; ZARZUELO, A.; VARGAS, F.; DUARTE, J. Effects of chronic treatment with quercetin in experimental renovascular hypertension. **Molecular Cell Biochemistry**, v. 270, n. 1, p. 147-155, 2005.
FANG, X. K.; GAO, J.; ZHU, D. N. Kaempferol and quercetin isolated from *Euonymus alatus* improve glucose uptake of 3T3-L1 cells without adipogenesis activity, **Life Sciences**, v. 82, n. 11, p. 615-622, 2008.
FERREIRA, D. F. **SISVAR:** software version 4.6 (build 61). Lavras: UFLA, 2003. Available at: <htpp://www.dex.ufla.br/danielff/dff02.htm>. Accessed on: January 15, 2013.
FIGUEROA-VALVERDE, L.; DÍAZ-CEDILLO, F.; CAMACHO-LUIS, A. Efectos inducidos por *Ruta graveolens* L., *Cnidoscolus chayamansa* McVaugh y *Citrus aurantium* L. sobre los níveis de glucosa, colesterol y triacilglicéridos en un modelo de rata diabética. **Revista Brasileira de Farmacognosia**, v. 19, n. 4, p. 898-907, 2009.
HENRY, J. A.; PANDIT, A. Perspective on biomaterials used in the surgical treatment of morbid obesity. **Obesity Reviews**, v. 10, n. 3, p. 324-332, 2009.
HSU, C. L.; YEN, G. C. Induction of cell apoptosis in 3T3-L1 preadipocytes by flavonoids is associated with their antioxidant activity. **Molecular Nutrition and Food Research**, v. 50, n. 11, p. 1072-1079, 2006.
KANEKO, J. J. **Clinical biochemistry of domestic animals**. 4.ed. San Diego: Academic, 1989, 932 p.
KÜHNAU, J. Flavonoids. A class of semi-essential food components: their role in human nutrition. **World Review Nutrition Dietetics Home**, v. 24, p. 117-191, 1976.
KWON, Y. I.; APOSTOLIDIS, E.; SHETTY, K. Inhibitory potential of wine and tea against a-amylase and a-glucosidase for the control of hyperglycemia associated with type 2 diabetes. **Journal Food Biochemistry**, v. 32, n. 1, p. 15-31, 2006.
LEVIN, B. E.; TRISCARI, E.; SULLIVAN, A. C. Metabolic features of diet-induced obesity without hyperphagia in young rats. **American Journal Physiology Regulatory Integrative Comparative Physiology**, v. 251, p. 433-440, 1986.
LIMA, A. R.; BARBOSA, V. C.; SANTOSFILHO, GOUVEA, C. M.C. P. Avalia^ao *in vitro* da atividade antioxidante hidroalcoólico de folhas de bardana. **Revista Brasileira de Farmacognosia**, v. 16, n. 4, p. 531-536, 2006.
MACHHA, A.; ACHIKE, F. I.; MUSTAFA, A. M.; MUSTAFA, M. R. Quercetin, a flavonoid antioxidant, modulates the bioavailability of endothelium-derived nitric oxide in diabetic rat aortas, **Nitric Oxide,** v. 16, n. 4, p. 442-447, 2007.
MAYER, M. A.; HOCHT, C.; PUYO, A.; TAIARA, C. A. Recent advances in obesity pharmacotherapy. **Current Clinical Pharmacology**, v. 4, n. 1, p. 53-61, 2009.
MORAES, L. D.; SOUSA, O. V. Qualitative and Quantitative Assessments of the Variation of Secondary Metabolites in *Tournefortia paniculata* Cham (Boraginaceae). **Revista Brasileira de Biociências**, *v.* 5, n. 2, p. 1032-1034, 2007.
MOTTA, V. M. **Biochímica clínica para o laboratório: princípios e interpretares.** 5. ed. Rio de Janeiro: MedBook, 2009, 400 p.
NOVELLI, E. L. B.; DINIZ, Y. S.; GALHARDI, C. M.; EBAID, G. M. X.; RODRIGUES, H. G.; MANI, F.; FERNANDES, A. A. H.; CICOGNA, A. C.; NOVELLI-FILHO, J. L. V. B. Anthropometric parameters and obesity markers in rats. **Animais de Laboratório**, v. 41, n.1, p. 111-119, 2007.
NASCIMENTO, A. F.; SUGIZAKI, M. M.; LEOPOLDO, A. S.; LIMA- LEOPOLDO, A. P.;

NOGUEIRA, C. R.; NOVELLI, E. L. B.; PADOVANI, C. R.; CICOGNA, A. C. Probability of misclassification as obese or lean in hypercaloric and normocaloric diet. **Biological Research**, v. 41, n. 3, p. 253-259, 2008.
OTA, M.; OKAMOTO, T.; HOSHINO, W.; WAKABAYASHI, H. Action of a-D-glucosidase from *Aspergillus niger on* dextrin and starch. **Carbohydrate Polymers**, v. 78, n. 2, p. 287-291, 2009.
PARK, M. Y.; LEE, K. S.; SUNG, M. K. Effects of dietary mulberry, Korean red ginseng and banaba on glucose homeostasis in relation to mRNA expressions of PPAR-a, PPAR-y and LPL. **Elsevier**, v. 77, n. 26, p. 3344-3354, 2005.
PATEL, C.; DADHANIYA, P.; HINGORANI, L.; SONI, M. G. Safety evaluation of pomegranate fruit extract: acute and subchronic toxicity studies. **Food and Chemical Toxicology**, v. 46, n. 8, p. 2728-2735, 2008.
PEREIRA, L. L. S.; MARCUSSI, S.; SÁTIRO, L. C.; PEREIRA, C. A.; ANDRADE, L. F.; DAVIDE, L. C.; SANTOS, C. D. Application of the Comet assay to evaluate the effects of white bean flour on the DNA of human lymphocytes. **Revista Brasileira de Ciências Farmacêuticas**, v. 48, n. 1, p. 103-108, 2012.
POPKIN, B. M.; GORDON-LARSEN, P. The nutrition transition: worldwide obesity dynamics and their determinants. **International Journal of Obesity**, v. 28, n. 3, p. 2-9, 2004.
RAYALAM, S.; DELLA-FERA, M. A.; BAILE, C. A. Phytochemicals and regulation of the adipocyte life cycle. **Journal Nutritional Biochemistry**, v. 19, n .11, p. 717-726, 2008.
RICARDON, K. F. S., OLIVEIRA, T. T.; NAGEM, T. J.; PINTO, A. S.; OLIVEIRA, M. G. A.; SOARES, J. F. Efeito dos Flavonóides Morina; Quercetina e Ácido Nicotínico no Metabolismo Lipídico de Ratos Alimentados Experimentalmente com Triton. **Arquivos Brasileiros de Biologia e Tecnologia**, v. 44, n. 3, p. 263 - 267, 2001.
SCALBERT, A.; JOHNSON, I. T.; SALTMARSH, M. Polyphenols: antioxidants and beyond. **American Journal of Clinical Nutrition**, v. 81, n. 1, p. 215-217, 2005.
SIMÁO, A. A.; CORREA, A. D.; CHAGAS, P. M. B. Inhibition of digestive enzymes by aqueous extracts of medicinal plants used in the treatment of obesity. **Journal of Medicinal Plants Research**, v. 6, n. 47, p. 5826-5830, 2012.
SIMÁO, A. A.; LAGE, F. F.; CHAGAS, P. M. B.; FRAGUAS, R. M.; FREIRE, J. M.; MARQUES, T. R.; CORREA, A. D. Antioxidants from Medicinal Plants Used in the Treatment of Obesity. **European Journal of Medicinal Plants**, v. 3, n. 3, p. 429-443, 2013.
SINGH, N. P.; MC
COY, M. T.; TICE, R. R.; SCHEIDER, E. L. A simple technique for quantifying low levels of DNA damage in individual cells. **Experimental Cell Research**, v. 175, n. 1, p.184-191, 1988.
SOUZA, S. P.; PEREIRA, L. L. S.; SOUZA, A. A.; SOUZA, R. V.; SANTOS, C. D. Study of the anti-obesity activity of the methanolic extract of *Baccharis trimera* (Less.) DC, **Revista. Brasileira de Farmácia**, v. 93, n.1, p. 27-32, 2012.
STRANGE, V. S., GOMES, T. D. U. H., ANDRADE, M. A.; BATITUCCI, M. C. P. Evaluation of the mutagenic effect of the crude hydroalcoholic extract, by means of *in vivo* bioassays and phytochemical prospection of *Cecropia glaziovii* Sneth (embauba), Cecropiaceae. **Revista Brasileira de Farmacognosia**, v. 19, n. 2, p. 637-642, 2009.
TUCCI, S. A.; BOYLAND, E. J.; HALFORD, J. C. G. The role of lipid and carbohydrate digestive enzyme inhibitors in the management of obesity: a review of current and emerging therapeutic agents. **Diabetes, Metabolic Syndrome and Obesity: Targets and Therapy,** Manchester, v. 10, n. 3, p. 125-143, 2010.
URZEDA, M. A.; MARCUSSI, S.; PEREIRA, L. L. S.; FRANCA, S. C.; PEREIRA, P. S.; SILVA, S.; GUIMARAES, C. L. S.; CALDERON, L. A.; STABELI, R. G.; SOARES, A. M.; COUTO, L. B. Evaluation of the hypoglycemic properties of the aqueous extract of *Anacardium humile*. **Evidence-Based Complementary and Alternative Medicine**, v. 2013, 8 pages, 2013.
VILLAREAL, D. T.; APOVIAN, C. M.; KUSHNER, R. F.; KLEIN, F. Obesity in older adults: technical review and position statement of the American Society for Nutrition and NAASO, The Obesity Society. **American Journal of Clinical Nutrition**, v. 82, n. 5, p. 923-934, 2005.
YAMAMOTO, Y.; OUE, E. Antihypertensive effect of quercetin in rats fed a high-fat, high-sucrose diet. **Bioscience Biotechnology Biochemistry**, v. 70, n. 4, p. 933-939, 2006.
WAN-IBRAHIM, W. I.; SIDIK, K.; KUPPUSAMY, U. R. A high level of antioxidants in edible plants is associated with genotoxic properties. **Food Chemistry**, v. 122, n. 4, p. 1139-1144, 2010.

WINTERBOURN, C. C.; GUTTERDGE, J. M.; HALLIWELL, B. Doxorubicin-dependent lipid peroxidation at low partial pressures of O2. **Journal Free Radical Biology and Medicine**, v. 2, n. 1, p. 1119-1122, 1985.

MIX
Papier aus verantwortungsvollen Quellen
Paper from responsible sources
FSC® C105338

Printed by Books on Demand GmbH, Norderstedt / Germany